BREAKOUT

BREAKOUT

Pioneers of the Future,
Prison Guards of the Past,
and the Epic Battle
That Will Decide America's Fate

NEWT GINGRICH

WITH ROSS WORTHINGTON

Library of Congress Control Number 2014381559
ISBN 978-1-62157-021-9 (hardcover)
ISBN 978-1-62157-281-7 (paperback)
First paperback edition published in 2014

Published in the United States by
Regnery Publishing
A Salem Communications Company
300 New Jersey Avenue NW
Washington, DC 20001
www.Regnery.com

Manufactured in the United States of America

10 9 8 7 6 5 4 3 2 1

Distributed to the trade by
Perseus Distribution
250 West 57th Street
New York, NY 10107

To the pioneers of the future,
who create so many better opportunities
for all Americans.

CONTENTS

PREFACE TO THE PAPERBACK EDITION ix

INTRODUCTION 1

CHAPTER ONE: Breakout: The Great Opportunity. 5

CHAPTER TWO: Breakout in Learning 25

CHAPTER THREE: Breakout in Health 49

CHAPTER FOUR: Breakout in American Energy 71

CHAPTER FIVE: The Green Prison Guards 85

CHAPTER SIX: Breakout in Transportation 107

CHAPTER SEVEN: Breakout in Space 125

CHAPTER EIGHT: Breakdown in Government 137

CHAPTER NINE: Breakout in Government 153

CHAPTER TEN: Breakout from Poverty 169

CHAPTER ELEVEN: Breakout in Achieving Cures 187

CHAPTER TWELVE: Breakout from Disabilities to Capabilities. . . . 193

CHAPTER THIRTEEN: Breakout Champions 201

CONCLUSION 207

ACKNOWLEDGMENTS 211

NOTES 217

INDEX 249

PREFACE TO THE PAPERBACK EDITION

Since *Breakout* was first published, two events have crystallized the core choices it presents. The first became a national scandal and led to the resignation of a cabinet secretary. The second was a private conversation with the former CEO of one of America's technology giants.

In the spring of 2014, a series of press reports detailed systemic corruption in the Department of Veterans Affairs. The alleged misconduct was outrageous: The destruction in Los Angeles of veterans' medical records to mask a backlog of appointments stretching months and in some cases years. A fraudulent scheme by senior VA officials in Phoenix to falsify data on wait times. The manipulation of records in Temple, Texas, to conceal long delays that threatened officials' bonuses. An investigation by the Office of the Inspector General uncovered schemes at 70 percent of major VA facilities to hide the unconscionably long waits that veterans had to endure before getting the medical attention their country owes them. The press reported that dozens of veterans died before they could see a doctor.

In other cases, it appeared that VA employees and even directors had engaged in outright theft. One former VA medical center director in Ohio pleaded guilty to "money laundering, wire fraud, mail fraud and conspiring to defraud the VA through bribery and kickback schemes in which he accepted tens of thousands of dollars from contractors in exchange for inside information."[1] The inspector general documented a number of other major thefts.

Reporting on the pockets of criminal behavior at the VA drew attention to the broader incompetence of the bureaucracy, which in all cases enabled the corruption and in many cases actually precipitated it. The directors ordered employees to create secret wait lists, for instance, because the centers were incapable of seeing patients within the required time frames. In other words, the bureaucracy built a system so dysfunctional that it required people—the vast majority of whom were surely well meaning, and in some cases veterans themselves—to break the law routinely or lose their jobs.

Actually, it's hard to say which is more scandalous: the islands of sophisticated corruption at the VA or the grand scale of general incompetence pervading the whole system. In an age when you can get cash from an ATM almost anywhere in the world within seconds, it takes on average 175 days to transfer a service member's medical records from the Defense Department to the Department of Veterans Affairs. To address this problem, the two agencies spent $1.3 billion between them to build a system of shared electronic medical records. But in February 2013, the two secretaries announced they were abandoning the effort because the new system was so plagued with problems.

No one could credibly argue (though many have tried) that these failures are for lack of manpower or funding. Between 2006 and 2013, the VA added nearly a hundred thousand employees, a 40 percent increase. Its budget increased even more—by 90 percent. So it is failing without the traditional excuses.

Instead, it's clear that the department is failing for systemic reasons that extend far beyond the shortcomings of any particular team of administrators. These systemic flaws are not limited to the VA. The same general incompetence punctuated by pockets of corruption infects the entire bureaucracy.

The problems at the VA escalated to a national scandal in part because they involved the mistreatment of our country's veterans, but in part, too, because Americans credited the reports immediately, recognizing them as only the latest in a long pattern of bureaucratic abuse and incompetence.

From the government's failure to build a functioning Healthcare.gov website within five years (a platform on which millions of people's health insurance coverage depended), to the IRS's targeting Americans for extra scrutiny on the basis of their political beliefs, to the routine indignities we suffer at our local DMVs, our daily lives now afford ample evidence that the bureaucracy is completely out of control.

The New Deal model of government that has predominated since the 1930s is failing before our eyes. The professionalized, centralized bureaucracy—which has grown for three-quarters of a century regardless of who was in power—is now unsuited to the modern world. It can't keep up with the astonishing technological progress we have seen in recent decades, some of which (like the internet) tends increasingly to expose the system's inadequacy and corruption.

The bureaucracy is failing, to be sure, but that does not mean it is going away. On the contrary, it is trying to clamp down. In many ways, today's bureaucracy has more power than ever before. Replacing this New Deal model will require an enormous fight. But it's not immediately obvious *what* the replacement should be.

That's why a recent conversation with Carly Fiorina, the former CEO of Hewlett-Packard, was the second event that clarified, in my mind, how much is at stake in the struggle I describe in this book—the struggle between the pioneers of the future and the prison guards of the past.

Carly was visiting our offices to offer some free advice from an accomplished leader on how to push forward a few of the ideas contained in the following chapters. All of her thoughts were insightful, but when she formulated the thesis of this book in her own words, she delivered a real epiphany.

The things that work in the twenty-first century, Carly said, will have four characteristics. They will be *digital*, *mobile*, *virtual*, and *personal*. It would be difficult to conceive of four words that more precisely describe what bureaucracies like the VA *are not*, and what all of the innovations we see cropping up around us *are*.

This, I realized, was the standard for the replacement model. If the replacement for the failing bureaucracy isn't digital, mobile, virtual, and personal, or if it simply tinkers with a broken machine, it too will fail.

An original purpose of this book was to argue that futuristic replacements for the failing old orders in education, health, government, and other areas were already emerging but were encountering the fierce resistance of those who wanted to use the power of government to protect the status quo. That conflict between the future and the past made the potential of innovation very much a political concern.

That is why we started an organization called Breakout America to serve as a center for every American who wants to accelerate this change. The potential for a politics of breakout is greater today than ever. Since the publication of this book, the contrast has only sharpened between innovations that are solving problems of everyday life and the bureaucracies that are making them worse.

Everywhere we look, we see rapid progress in many important fields while the competence of government deteriorates. We may have hundreds of thousands of veterans waiting to see doctors at the VA, for instance, but we also have Theranos, a company in California that has developed the technology to conduct one thousand of the most common medical lab tests automatically, using very small blood samples and for less than half of their current prices. Theranos machines, which could soon be available in every Walgreens drug store, would email you the results of your test by the time you walked out the door. This is a digital, mobile, virtual, and personal approach to healthcare—one that Americans would embrace voluntarily rather than needing bureaucracies to force it on them.

Although the details may change, I suspect the stories in this book will be relevant for years to come. The conflict will go on until we make the political choice to break out from the prison guards of the past.

Every year that the prison guards have their way, every year that they manage to hold back the future, Americans suffer. The cures we don't achieve, the students we don't teach, and the opportunities we don't seize are the tragic if hidden cost of a government commandeered to block the future. I am convinced that we must break out, and we must do it soon. And with your help, we will.

—Newt Gingrich

INTRODUCTION

Even before the painful sore appeared in her mouth just before Christmas in 1999, Abigail Burroughs was an exceptional young woman. She had earned awards for her community service in her hometown of Falls Church, Virginia, and had volunteered in homeless shelters. She had graduated second in her high school class, and now she was excelling in an honors program at the University of Virginia.

Of all the reasons we might have read about Abigail in the *Washington Post*, a desperate battle with bureaucracy over access to a lifesaving drug would have seemed the least likely. It was a cruel battle she shouldn't have had to fight.

"It was annoying," Abigail told the *Cavalier Daily*, her college newspaper, referring to the sore in her mouth. "I just wanted it to go away." She finally convinced her doctor to remove it. Before classes resumed after winter break, Abigail heard from the doctor. The lab results were shocking: the sore was cancerous. She was nineteen years old.[1]

Still, Abigail's prognosis was good once the surrounding tissues were removed. Her doctor told her there was a 90-percent chance the cancer wouldn't come back. The short procedure should have been the end of the matter.

But it wasn't. The cancer returned a few months later, this time as a larger lump in her neck. She had it removed surgically and quickly began radiation and chemotherapy, which left "severe burns on the inside of her mouth."[2] But Abigail was optimistic she would recover and return to school, where she had a boyfriend and lived in an apartment with other girls her age.

Not long after the first round of treatment, though, her doctors found another lump in her neck. They discovered the cancer had spread to her lungs and stomach too. The doctors in Charlottesville had nothing else to try. But doctors at Johns Hopkins, one of the leading cancer centers in the country, knew about two breakthrough drugs that showed promise for cases like Abigail's. They were her best chance, the doctors said.

Unfortunately, neither drug was approved by the Food and Drug Administration, the federal agency that must clear virtually every medical product for safety and effectiveness. Both drugs had passed the early stages of trials, indicating that they were safe for use in patients. But it would take many years—and hundreds of millions of dollars—for the companies behind them to prove to the FDA's highest standards of scientific evidence what some of the best doctors in the field had already observed: these drugs could help cancer patients who had failed to respond to the traditional treatments. Abigail Burroughs didn't have years to wait for definitive evidence while the FDA haggled with the drugmakers. She was dying. She needed the breakthrough treatments immediately.

Abigail's doctors tried to enroll her in the clinical trials the companies were conducting to obtain FDA approval. Either of these would have given her a chance of getting the real treatment—or of getting a placebo (which of course meant certain death). But Abigail's condition didn't meet the exacting standards for even this undesirable choice.

The FDA offers just one recourse for patients who urgently need access to breakthrough drugs if they don't make the clinical trials. Patients can ask the drugmakers for "compassionate use," which, if granted, must be approved by FDA bureaucrats on a case-by-case basis.

At this point Abigail's story began to get national attention. It started at the University of Virginia, where the administration, fraternities, and student groups organized a campaign to petition the companies to give Abigail the drugs. Then the campaign spread. Thousands of people she had never met began writing to the drug companies on her behalf.

At the same time, her father founded the Abigail Alliance for Better Access to Developmental Drugs to push for wider access to new drugs for terminally ill patients who had no other options. He argued that it is unjustifiable to keep the drugs from dying patients, even if they carry risks.

A reporter from the *Washington Post* visited Abigail in her apartment and published a heartbreaking story on her struggle. He reported that the college junior "rates her life by what she can do—a daily visit to her favorite Starbucks—or how far she can walk—she slept 24 hours to gather strength to make it to a Dave Matthews concert last month."[3]

Despite months of public pressure on Abigail's behalf, the companies refused to give her the drugs. It seemed that in the middle of their billion-dollar effort to win FDA approval, they were unwilling to risk a bright twenty-one-year-old girl's dying while receiving the treatment that was under scrutiny. For Abigail, it would be the FDA-sanctioned treatments, or nothing at all.

"I try not to think about [the companies], because when I start to, it makes me really, really angry," she told the *Post*. "I can't understand how these people can be so nonchalantly by-the-book and just say no."

Abigail passed away about a month later. Her father carried his quest for more open access all the way to the Supreme Court, which declined to hear his challenge to the existing system. His suit attempted to make available one of the drugs to which Abigail had sought access, C225, known today as Erbitux.

Researchers invented Erbitux in 1983, when Abigail was three years old.[4] Only a small fraction of drugs passes successfully through the regulatory gauntlet, and it took a decade before they found a company willing to bear the tremendously expensive risk.[5]

By the time Abigail asked for Erbitux in 2001, eighteen years after its invention, it had already shown promising results. Yet it did not receive FDA approval for three more years, and then only for colorectal cancer. It took until 2006 to win approval for localized head and neck cancer. The drug was not approved for late-stage head and neck cancer like Abigail's until 2011, ten years after her death, eighteen years after a drug company acquired the rights, and twenty-eight years after it was invented.

Abigail's story is tragic, but it is the norm. Somebody was willing to lock her in the past, denying her the breakthrough drug of tomorrow that might have saved her life. And as you will read here, the forces fighting "to keep the Past upon its throne" are not limited to medicine. They are at work in many important fields.

This is what we must change.

CHAPTER ONE

BREAKOUT

THE GREAT OPPORTUNITY

America is on the edge of a breakout.

Astonishing progress in medicine, transportation, learning, energy production, and other areas has set the stage for one of the most spectacular leaps in human wellbeing in history. But like Abigail Burroughs, we might never reach the breakout these breakthroughs make possible.

Even as pioneering scientists, engineers, and entrepreneurs apply their genius and industry to overcoming some of our most serious problems, there are institutions, interests, and individuals hard at work to thwart this breakout. Their power and comfort are often bound up with the status quo. They are willing to forgo the breakout, to keep us prisoners of the past, in order to cling to the privileges that the old order has bestowed on them. And they will succeed if we let them.

This book is about that contest. You will read about the almost unbelievable advances that are on the threshold of reality—triumphs of human ingenuity that will save millions of lives and vanquish afflictions that have vexed us for centuries. And you will meet those who are trying to kill that breakout before it begins.

The defining battle of our time is not between the Left and the Right. It is between the past and the future. And every American is a contestant, whether he or she knows it or not. As you read in the following chapters about breakthroughs in one field after another, you will see that the choice before us could hardly be more urgent.

Breakouts Past

Breakouts are bigger and more powerful than breakthroughs. In a dynamic, entrepreneurial society like America, breakthroughs are happening all the time in various fields. They are quickly absorbed into our economic and social routine.

There are times, though, when a combination of science, technology, innovation, and entrepreneurship creates waves of new possibilities that reinforce each other. Then, in a matter of thirty or forty years, the world is dramatically different.

Breakouts change how we live and how we think about living, how we organize activities and how we organize government.

To understand the magnitude of change in a breakout, consider the world that emerged in the decades just prior to 1870. In only a generation, the steam engine had revolutionized the economy, drawing millions of Americans from lives of subsistence farming to work at factories in major cities. Sewing machines had radically changed the way people made clothes, cotton gins had sped the production of cloth, and motorized farm equipment had transformed agriculture.

Steamboats traveled up and down American rivers routinely and had shortened the journey across the Atlantic from more than ten weeks by sail to just twelve days by steam. The transcontinental railroad, also powered by steam, had enabled fast long-distance travel and made shipping goods across the United States practicable.

The camera had produced the first photographs, and the telegraph had enabled simple communication nearly at the speed of light. Networks had emerged to carry news throughout the United States and Europe in minutes rather than weeks.

If life in 1870 wasn't what we would call comfortable, it was materially much better and technologically much more advanced than it had been in 1840. To a person in middle age, the world was a marvel of technological achievement.

He might read Jules Verne's *Around the World in Eighty Days* (the speed was intended to be impressive), published in 1873, and imagine an incredible future.

He would also worry about the uncertain new world. Cities were bursting with people. In 1820 there were only five U.S. cities with a population over twenty-five thousand; by 1850 there were twenty-six.[1] They were crowded and unpleasant. Disease was rampant. Immigrants, drawn by the new opportunities, flooded in from all over the world, to the dismay of many Americans.

No one could have imagined the world that was just around the corner. The impressive technological progress that Americans surveyed in 1870 only set the stage for what was to come. Candles still lit homes at night, ice had to be cut from lakes and stored in iceboxes, and most travel was by foot or horse.

In the lifetime of someone born in 1870, Edison invented the electric light bulb, which quickly replaced candles as the dominant source of light. Refrigerators replaced iceboxes, washing machines replaced washboards, and vacuum cleaners replaced many brooms. Electric fans let people cool themselves with artificial wind.

The telephone enabled distance communication by *voice*, and radio enabled a new kind of mass communication completely different from newspapers. For the first time, any citizen could hear the words of prominent Americans, and everyone could listen to the same music and live sporting events. Gramophones and motion pictures became available, replacing vaudeville as forms of popular entertainment. That common culture helped assimilate the millions of new Americans who had arrived on the steamships.

The internal combustion engine was developed, and the automobile replaced the horse and buggy. Americans now enjoyed flexibility in where they lived and worked, and Detroit became a boomtown. In cities, underground transit systems—subways—connected far-flung neighborhoods. The Wright brothers made their first flight in 1903, and in 1927, just fifty-four years after Verne's *Around the World in Eighty Days*, Charles Lindbergh flew from New York to Paris in thirty-three hours.

Physicians began using X-rays and the newly invented electrocardiogram. In the early 1920s, scientists identified vitamins, and in 1928 they discovered penicillin.

By 1930, pioneering Americans had created some early version of virtually everything we think of as modern.

The generation that lived through that change found itself rethinking work, leisure, retirement, and government. The reform movement that created much of modern government was the political parallel of the explosion of new science, technology, and entrepreneurship.

Until the Great Depression began in 1929 (and, for most Americans, even after it began as well), life was simply much better than it had been in 1870. Life was better not because politicians had the right plans or because government built a bridge to the future, but because millions of Americans embraced the opportunities to improve their lives. They solved many of the problems of daily life and transformed it in the process. They had produced a real American breakout.

Breakouts Future

Americans in the second decade of the twenty-first century have witnessed a breakout in the field of information. Soon after Motorola made the first handheld mobile telephone in 1973, Apple and Microsoft opened the first act of the information revolution with the personal computer. The next act began in the early 1980s as the internet began to emerge. By the mid-1990s, it was widely available, and we got AOL, Amazon, Yahoo!, and, in 1998, Google.

By the mid to late 2000s, we were into the third act, when the internet became a major part of our lives. A high-spirited Harvard undergraduate, Mark Zuckerberg, launched a campus photo site called

"Facemash" in 2003, whose instant popularity crashed the university's network. In 2004 he launched Facebook, which nine years later had more than a billion users. This deep integration of information into our lives continued with ubiquitous iPhones and Android devices constantly connecting us to the virtual world. In 2013, Google reported that 1.5 million new Android devices are activated every single day.[2]

If you have lived through these changes, you know how different the world is today from your childhood. Imagine having no personal computer, no cell phone, no Google, no Wikipedia, no Facebook. Imagine a world with no online shopping, no online hotel and airline reservations, no Google Maps. Without their iPhones, many people today are lost, literally.

Our desperation when our batteries die—we race to find a plug like scuba divers whose tanks have run out of air—is a reminder of how much we have woven our lives around a two-hundred-dollar device that no amount money could have purchased ten years ago.

It is all quite amazing. But like people in the 1870s marveling at steam engines and railroads, we have not quite digested the change that has taken place or begun to understand the change that is to come as breakthroughs continue in other fields.

We have seen an information revolution, but there have been no comparable advances in healthcare, education, transportation, manufacturing, or government.

What's the holdup? After thirty years of exponential improvement in computers, why does education still look the same—or worse, and more expensive? Why is healthcare still fundamentally the same—or worse, and more expensive? Why is government disastrously worse and *much* more expensive? Why, outside the confines of our pockets and our computer screens, is everything so old?

The fact is, we got lucky with computers, and especially with the internet. No one succeeded in blocking their future, largely because both were open systems. The big research labs of the 1970s at places like IBM and AT&T didn't have a veto over the emerging personal computer industry, because before long, anyone with a little know-how could assemble the components in his garage and start tinkering around with code. If he was innovative enough, his ideas could catch on. And they did: two of those garage tinkerers founded Apple and Microsoft.

Even as the personal computer industry matured, it remained remarkably open to innovation. No one had the authority to tell Steve Jobs or Bill Gates the one way he had to make his computers. The technology could advance just as fast as engineers could build faster computers and software developers could come up with new ideas to program. Everyone else quickly imitated the good ones.

Need proof that no one, no matter how rich or powerful, had a veto over progress in computers? In 1998 the U.S. government brought an antitrust case against Microsoft because it so thoroughly dominated the industry. Apple's market share was just 4 percent at that time. Today, Microsoft's entries in the smartphone and tablet markets are on the verge of irrelevance, in a distant third place behind Apple and a company that was still operating out of a garage in 1998—Google. In the wide-open world of consumer electronics, no one's dominance is secure for long. Challengers can emerge on the stage before the leaders know what hit them.

The internet has thrived on the same radical openness to innovation. Anyone—Microsoft or a kid in Bangalore—can plug in and offer ideas, tools, services, or musings to everyone else. The barriers to entry couldn't be lower. Have some interesting photos? Millions of people want to stop by and see them. Have a better web browser? Overnight, it has a hundred million users. Have an idea to revolutionize the way friends connect? No one has the power to stop you.

It is because these systems are so open that improvements spread rapidly. Consumer electronics companies, web startups, and individual programmers all over the world actually compete to keep you ever more pleased with your digital experience. They're in a race to see who can make the next big thing better, cheaper, faster, smarter, and cooler. And they're in that race because nobody—no matter how big—has the power to push "Stop" on the race and say, "I've won. I think we'll just keep things as they are for a while." Some would like to. But no one has the authority to preserve the status quo.

This openness to innovation is what's so amazing about the internet. It's why we have iPhones and iPads and Google and Twitter and YouTube and everything else we love to spend hours poking away at.

And it's what is missing in the other areas of our lives where we have seen no such improvement. In education, in health, in transportation, and in government, the old order has a veto. There are people empowered to protect the past, to say, "This is the best things are going to get for a while." You probably know who some of them are. The others, you're about to meet.

We are on the edge of an era of change the likes of which we have not seen since the period between 1870 and 1930. This astounding breakout could transform our lives in ways we can only begin to imagine today. It could all happen within our lifetimes—or not. The breakout first has to make it past the guardians of the old order, who will be only too happy to smother it.

It will take all of us—and some very big fights—to make sure that does not happen.

The next breakout could transform education, manufacturing, medicine, energy production, transportation, and government as thoroughly as the internet has transformed communication. Of course, we cannot know exactly what this new world would look like, but we can already begin to see its outline taking shape.

Education

Education pioneers are using the information revolution to develop new ways of teaching that will tailor lessons to each individual student. We could abandon the old "butts in seats" system, which measures progress by the number of hours students sit in chairs listening to teachers. The one-size-fits-all system is failing because it educates the "average student." Instruction is too slow for one half of the class and too fast for the other.

In higher education, despite the outrageous prices, the model is even more hostile to students. Too often, it goes something like this: "I am the professor and I already know the material, so you will listen to my lecture and then figure it out for yourself if you don't understand."

Only the most unusually motivated students thrive throughout seventeen years of this experience.

The explosion in computing and networking technology, however, allows us to replace this failing system with one that can adapt to every student. An achievement model powered by computers and informed by brain science and psychology can teach students exactly what they need, as they need it, in the way that is optimal for each unique, individual learner.

Almost as important as the improvements in quality will be the sharp cuts in cost. The new system should cost only a fraction of what we spend on today's dysfunctional education bureaucracy. Imagine a high-quality basic education that is free to everyone in the world. Imagine ten-thousand-dollar bachelor's degrees from America's best universities. We are talking about a big leap in opportunity for millions of people whom the current system serves so poorly.

Materials and Manufacturing

Only a few years ago, three-dimensional printers sounded like something out of science fiction. Now, these devices are sold at Staples. Like inkjet printers, 3-D printers allow consumers to "print" designs they've created on or downloaded to their desktop computers. Instead of laying ink on a sheet of paper, however, 3-D printers lay down thin layers of plastic (or some other material), one on top of the other, in order to build an object.

Because the technology is still in its primitive stage, early adopters are figuring out new uses for it all the time. Entrepreneurs who once sent designs to China and waited for models to arrive by mail are using 3-D printers to prototype their products quickly and inexpensively. Schools and universities are uploading digital models of fossils and bone specimens to the internet so amateur paleontologists anywhere in the world can print out a copy of a mosasaur tooth at scale. Children are using them to print new tokens for their board games and figures for their dollhouses. Second Amendment proponents are using them to create firearms that can be transmitted over the internet. And doctors are using advanced 3-D printers filled with an "ink" of *living cells* to print new human organs.

Soon you should be able to "print" in your home almost any physical object you can imagine. The possibilities for manufacturing, health, education, and other endeavors are amazing.

Other discoveries in materials science, like nanocarbon tubes, could change the world just as much as plastics did in the second half of the twentieth century.

Health

As the prospect of printing new human organs suggests, we are on the cusp of a transformation in healthcare. Now that consumer computers can process such massive amounts of data as the human genome, genetics is finally becoming a practical tool in medicine.

The implications are enormous. Just as we can replace teaching the "average student" with personalized learning, with the advances in genetics, we can replace treating the "average patient" with personalized medicine. Healthcare should be better, with fewer unintended side effects, and could cost less.

The most revolutionary effect of these breakthroughs could be the replacement, in many cases, of disease management with *cures*. In the United States, people no longer show up at the doctor with polio. We no longer treat tuberculosis or malaria. We've cured them. They're nonproblems.

Imagine a world where diabetes is cured, where Alzheimer's doesn't occur, where a cancer diagnosis means a night in the hospital while doctors install a new organ grown from your own cells. We could reach that world in the foreseeable future.

Energy

A decade ago, the airwaves were full of people who were certain the world was running out of oil. Americans must face rationing or punitive gasoline taxes, they said, to slow the oncoming crisis of "peak oil." National leaders cried hysterically for an energy policy to force the transition from this old, dirty, and dwindling sludge to solar, wind, and other underdeveloped technologies that would impose atrocious costs on the American people but would at least avert disaster.

No one thought of so-called "fossil" fuels as *the future.*

Then came what history will remember as one of the most significant technological breakthroughs of the twenty-first century: the union of "hydraulic fracturing"—"fracking" for short—and "horizontal drilling."

Like 3-D printing, fracking sounds like science fiction. Energy developers drill a well bore as deep as ten thousand feet. Then, incredibly, they force a ninety-degree turn in the steel pipe and drill horizontally for up to two miles. Next, they pump a mixture of water and sand at high pressure into the well, producing small fractures in the rock far underground. When they withdraw the water, the sand remains and keeps those fractures propped open, allowing *individual molecules* of trapped oil and natural gas to migrate up the well.

It would be an understatement to say this engineering marvel has caused a revolution in American energy. Because of fracking, the United States will soon be the world's leading producer of oil and natural gas.

Fracking has killed the peak oil myth we were fed for decades and with it, the crisis rationale for inflicting on Americans the pain of higher taxes or rationing. But more importantly, it could be a game changer for the American economy and national security.

Transportation

The cars we drive today are safer, more comfortable, and more efficient than the cars our parents and grandparents drove—but they are fundamentally the same. That could soon change. After decades with no groundbreaking innovation in automobiles, Google has developed genuinely self-driving cars that have traveled more than six hundred thousand miles on California roads virtually without incident.

After being beaten to the automotive future by an internet search company, most of the major carmakers are now racing to design self-driving cars of their own. Sometime in the third decade of the twenty-first century, cars requiring little if any human intervention could become commercially available. Already today, companies like Caterpillar are operating giant self-driving trucks in industrial settings.

Because roughly 80 percent of car accidents are the result of driver error,[3] there is good reason to think this breakthrough could save tens of thousands of lives and hundreds of billions of dollars. Yet improved safety will be only one consequence of the second great automotive revolution. Cars and trucks that drive themselves from place to place, ferrying passengers and cargo, would transform the way we live.

Government

All over the world, governments are having to adapt to the internet. In China, Egypt, Turkey, and countries throughout the Middle East, old regimes are faced with powerful new social forces they cannot control. The internet exposes the weakness of these regimes.

In the United States as well, the internet has begun to reveal the corruption of the bureaucratic state. Virtually all business is conducted through email, providing the American people with a detailed record of their public servants' conduct. The picture has not always been flattering (and in a number of cases, we have seen bureaucrats conspiring to break the law by using their personal email addresses to avoid leaving a trail).

New technologies have the potential not just to expose bad government but to improve it as well. Yet with so many astonishing changes happening all around us, the pace of innovation in government seems glacial. Our government and our politics are trapped in the past. But there are hints of the future we could achieve using technology to empower citizens to reclaim the functions of government from the bureaucratic state.

California's lieutenant governor, Gavin Newsom, has argued that citizens can use technology "to bypass government, ... to take matters into their own hands, to look to themselves for solving problems rather than asking the government to do things for them."[4]

Across the country we will see that there are fascinating experiments taking place at the local level to do just that. We have not even begun to understand how the information revolution we have just lived through—and the breakthroughs yet to come—could transform government and enhance freedom and prosperity for the next generation of Americans.

Enemies of the Future

These revolutionary breakthroughs in information, transportation, education, energy, materials and manufacturing, healthcare, and government are no longer mere daydreams. They're all beginning to happen right now.

Even those changes that seem like science fiction, however, are just the earliest stages of the world we could one day know. In many areas we have only vague but exciting indications of what is yet to come. We know as little about the future as someone living in the world of candles and horses knew about the age of television and passenger planes.

The change that is coming won't be simply more of the change we have seen in the last generation. It will be something else entirely—a change of kind, not just a change of degree.

We are talking about a fundamental transformation of what is possible, what we can accomplish, and what it will cost.

The scale of this transformation makes it a watershed. For it to happen, we must reorganize how we think and act, how we structure organizations, how we organize activities, the very questions we ask, and the metrics we establish.

Many of the problems we spend so much time and effort trying to solve today can be cured out of existence tomorrow. No tinkering around the edges compares with the breakout ahead.

If we achieve breakout, it will be one of the most momentous events in our history. It will bring more people out of poverty (several billion) than ever before. It will create more opportunities and more new products and services than we can count. It will solve budget issues that currently appear unsolvable. Smaller breakouts in healthcare and education will improve quality of life by an order of magnitude.

We are talking about a completely different world—like a jump to another century. It can all happen, and it can all happen soon. America can break out. But the prison guards of the past, the guardians of the old order on both the Left and the Right, will not go quietly into the night.

Think back to the shift from the candle to the electric light.

Candles had been around for about five thousand years. They provided some light, but for most of that time, people ended their days when nighttime began. Candles didn't offer enough light to do much of anything, and they were expensive. When you were paying to burn candles, even activities like reading were economic decisions. Young Abraham Lincoln's family was too poor to burn candles for something as frivolous as reading, so he learned to read by the light of the fireplace.

In 1870, gas lamps were available in some places, but they were basically candles with more fuel. The night was dark, the stars were bright, and no one could imagine things any other way.

Then in 1879, Thomas Edison introduced the first reliable electric light, made from carbonized bamboo thread. It would burn for 1,300 hours.

Edison himself understood the revolution that his invention represented. "After the electric light goes into general use," he said, "none but the extravagant will burn tallow candles." The poor—along with the rest of America—would be illuminated through technological change. The inexpensive light would improve reading and literacy, add more useful time to the day, and brighten homes and street corners.

The electric light is not simply a cheaper or better candle, however. It is a different thing entirely, and it opened up completely new possibilities over the next half century. Think of Times Square in New York or the Las Vegas Strip at night. Think of an airport runway. Think of theater lighting or a searchlight or a camera flash.

These things were new. Much like the transformational breakthroughs we are approaching in the twenty-first century, the change from the candle to the electric light was a change of kind, not a change of degree. And it happened suddenly, in a few decades.

Electric lights make life better, happier, and more prosperous, and yet because they are so cheap, even the poorest among us take them for granted.

Today we even have people who worry about "light pollution," a term that in 1870 might have conjured images of smoke and soot from a lone candle but that in fact represents the phenomenon of having so much artificial light that you cannot see the stars.

Light pollution activists want people to use dimmer bulbs, to turn off extra lights, and to stop using bulbs of certain hues.

Can you imagine explaining to someone in 1870 that in the next century, there would be groups organized to fight "light pollution"?

Now try to imagine that today's environmentalists were around when electric lighting was coming into general use. They would have declared the light bulb hazardous to nocturnal animals and a grave threat to the night sky.

Cynical columnists would have announced that electricity can kill and that the entire idea was an effort by profit-seeking corporations to exploit the poor and make them want something they didn't need.

If some of today's famous economists had been around, they would have joined forces with the candlestick makers' union and called for the protection of thousands of candlestick-making jobs by banning this dangerous new product, which was destroying American jobs by creating free light.[5]

If today's bureaucrats had been around, they would have established an agency to plan the distribution of electric lights and to set fixed standards for their design, since Edison and other inventors couldn't be trusted to promote the public interest.

Politicians would have announced there was an electric light shortage and that we needed a redistribution program so the poor would get their fair share of candles.

Fortunately, for most of America's history, such opposition to the emerging future was weak and lacking authority. Before World War II, there was no army of lobbyists, unions, bureaucrats, and litigators seeking to cancel the future and protect their own privileges.

The railroad did not have to battle a stagecoach union. The automobile did not have to get past the lobbyists of the horse and buggy industry. The electric light did not have to run a gauntlet of regulation and litigation.

If all these organized and entrenched interests had been around in the late nineteenth century, then not only the light bulb but the entire breakout that produced the modern world might never have happened, or at least have been considerably delayed.

The leaders of the twenty-first-century breakout find themselves acting in a historic drama that was unknown to their predecessors. We don't know yet if that drama will be a comedy—in the classical sense of having a happy ending—or a tragedy. What we do know is that every American has a role to play. Most of us are neither heroes nor villains in the breakout drama. But as citizens and voters, we are part of a large supporting cast. One reason I wrote this book is convince you how important those supporting roles are—for good or for ill. In the chapters that follow, as we explore the different facets of the next breakout, you will get to know the cast of characters in the great drama of our time, and you will have to choose your own role.

The Cast of Characters

There is nothing more difficult to carry out, nor more doubtful of success, nor more dangerous to handle, than to initiate a new order of things. For the reformer has enemies in all those who profit by the old order, and only lukewarm defenders in all of those who would profit by the new order ... [because of] the incredulity of mankind, who do not truly believe in anything new until they have had experience of it.

—Niccolò Machiavelli, *The Prince*

Although Machiavelli wrote these words five centuries ago, they describe the four main characters in the drama now playing out around us as well as they described the actors in the palace intrigues of Renaissance Florence.

The defining struggle for America in the next few decades is a political struggle, but it is a fight between the future and the past, not the Right and the Left. If the American people do not dethrone the protectors of the past, the breakout we could achieve will be delayed beyond our lifetimes and perhaps even killed.

The great struggle between those who are creating the future and those who want to keep us trapped in the prisons of the past is happening every day all around us, whether we realize it or not.

The Pioneers

Across America there are citizens pioneering the future, developing an amazing range of breakthroughs. They are creating opportunities for greater prosperity, more jobs, lower costs, more choices, healthier and longer lives, and greater national security. One at a time, they offer us pieces of a better world.

We know the names of some pioneers, men and women like Steve Jobs, Bill Gates, Sergey Brin, Larry Page, Sheryl Sandberg, and Marissa Mayer. Most pioneers of the future, however, are not famous. In fact, thousands of them are working on medical and biological research, developing apps that solve problems, engineering a future of energy abundance, and fighting to revolutionize education. You will meet a few of them in this book.

People with new ideas about how we can do something better or cheaper—or do something we never even thought of—change the world. They can lead breakout.

The Prison Guards

Wherever the future is happening, however, there are opponents who want to stop it—"those who profit by the old order," who protect themselves and their privileges. These are the prison guards of the past.

A breakthrough in learning? The teachers' unions and the mandarins of the academy will discredit it, insisting that they alone can be trusted with the welfare of our youth.

A breakthrough in energy? The extreme environmentalists will invoke an ecological catastrophe, accuse the pioneers of poisoning the wells, and try to regulate and litigate the breakthrough into oblivion.

A breakthrough in medicine? With a compliant administration in Washington, the Food and Drug Administration and the bureaucrats in both public and private insurance can make sure it never sees the light of day.

It has been said that the future has a publicist, but the past has lobbyists. And sometimes the future doesn't even have a publicist.

The prison guards—politically powerful, well funded, savvy, and brutally determined—have enjoyed a lot of success lately. Ask yourself why forty-five years after landing a man on the moon, the United States has no spacecraft capable of carrying men into orbit. Ask yourself why so many public schools never improve despite our spending more and more money on them, and why millions of families have no alternative to these hopeless holding pens. Ask yourself why a college education costs as much as buying a house in an age when information has become virtually free. Ask yourself why practically every doctor expects the next medical breakthroughs to be available in China and Europe before the United States.

Despite all the remarkable changes in information technology in recent years, the prison guards have kept us trapped in the past in most other fields. In fact, our fascination with high-tech consumer electronics may blind us to our prison bars. Our devices lend an appearance of the future to otherwise old ideas and institutions of education, medicine, government, and more.

The Prisoners

Those who are tricked into believing that these outmoded models are the best we can do are prisoners of the past—in Machiavelli's version, those "who do not truly believe in anything new until they have had experience of it." The ranks of the prisoners include most of the news media, most members of Congress (in both parties), and sadly, most people, period.

A dose of skepticism about bold promises is natural, of course, even healthy.

But prisoners think the only solutions to our current problems are more or less of the same solutions that are not working now. Believing our current challenges represent a new normal, they close their minds to the possibility of a better future.

If schools are performing badly, the prisoners debate spending more or less money on the same teachers teaching the same way.

If you have a medical condition that is debilitating or killing you, the prisoners smile sympathetically as they explain that you cannot have the most advanced treatment, because it has not been certified by the bureaucrats who will take fifteen years to approve it.

If government is spending far more than it takes in, the prisoners recoil from cutting back or cutting out any of the bureaucracy. The choice, they assume, is between raising taxes and cutting essential services that people depend on—and that's no choice at all.

The prison guards, who make it their business to ensure that the status quo is inviolable, have managed to convince the prisoners that this is the best we can do.

The prisoners are not the villains of this drama. They're the victims, trapped in bad schools, enduring a dysfunctional healthcare system, paying the unnecessarily high price for energy, and suffering under historic unemployment. The prisoners suffer while the prison guards profit.

The prisoners are, however, the enablers of their own captivity. The guards can succeed only with the consent of those who refuse to "truly believe in anything new until they have had experience of it." They are trapped in the mindset that we can't do better. Yet if we can break enough prisoners out of that mindset, the guards will soon be overwhelmed by popular demand.

The Breakout Champions

To awaken the prisoners from the spell of their guards is the task of the final group: the breakout champions, those who believe that we can break out. They are often surrounded by people trapped in the past, skeptical about any new idea. This discouraging environment leaves pioneers of the future with "only lukewarm defenders" instead of true champions.

It is up to the champions—to the American people, to us—to be not lukewarm defenders but real advocates. It is up to us to assert that this is not the best we can do, that there is a dramatically better future just ahead, and that we can overcome the prison guards of the past and break out.

If you want to be a breakout champion, this book is for you. In the coming pages, you will read a report from the frontiers of the future. You will see the prison guards—often the self-styled arbiters of "enlightened" opinion—desperately defending the old order and keeping America trapped in the past. Finally, you will learn what we together must do in order for America to break out.

The fight will be to modernize our institutions, our laws, and our regulations so we can see a genuine breakout in our lifetimes. That is the job of the breakout champions. That is the job I hope to convince you to take on.

CHAPTER TWO

BREAKOUT IN LEARNING

Caitlin Pierce is a normal teenager from Arkansas. In May 2013 she graduated from high school at the age of eighteen. The very next week, she graduated from college with a bachelor's degree.

Caitlin was homeschooled and had spent her junior and senior years of high school earning real college credit in courses coordinated through a Texas-based company called CollegePlus. Its average student earns a degree in two years at a cost of about fifteen thousand dollars.[1] Like her high school peers who will spend the next four years in college, Caitlin will begin her working life with a degree from an accredited institution. But unlike most of her friends, she won't be saddled with tens of thousands of dollars in debt. Caitlin will be better off because of the income she will earn during the four years her friends are in college and because of the money she won't spend over the next thirty years paying off student loans. CollegePlus offered Caitlin a win-win proposal.

Perhaps the most remarkable thing about Caitlin's achievement is that in 2013, her story is still unusual.

We are on the edge of a dramatic transformation from bureaucratic education to individualized learning. The technologies of communications, information, and learning are evolving so rapidly that they could soon overpower the prison guards of the past, who have been fighting desperately to sustain the education bureaucracy even as it fails to serve our children's and our country's needs.

In the last generation, almost everything about how we communicate information and knowledge has changed. The mail we receive from the postal service is most likely junk since important communication now happens by email. We have fewer reasons to go to the library now that we have Kindles and iPads for reading and Google for researching. We don't buy expensive encyclopedias because an infinitely richer online version, Wikipedia, is free and constantly updated. In fact, expensive encyclopedias are going out of business. Fewer and fewer of us subscribe to print newspapers because we generally read our news online or maybe just absorb it from tweets and Facebook posts.

Everything about transmitting knowledge has changed. Everything, that is, except our schools and universities. The cinderblock classrooms with thirty-five desks and nine-pound textbooks, spiral-bound notebooks and number 2 pencils are still much the same as they were in 1970, or, for that matter, in 1930.

It's true that most schools and universities now have computer labs, and some even provide student email addresses and class websites, but we have failed to fundamentally rethink learning in a world where information is increasingly free, searchable, and available on demand.

If anyone can learn to play the piano from an application on a tablet propped on the music stand, or take exercise classes on YouTube instead of going to the gym, or even learn online how to build a backyard deck, then why do students today still have to sit shoulder-to-shoulder, staring at chalkboards in stuffy classrooms in order to learn multiplication?

If learning on demand is becoming the norm in the real world, why is education on schedule still the bureaucratic norm?

The prison guards of the past who lead the opposition to change from education to learning have forestalled the kind of transformation we've seen in every other information industry.

Despite the layers of prison guards (teachers' unions, education bureaucracies, state curriculum rules, etc.), a few pioneers are beginning to break through the resistance.

Khan Academy

Ten years ago Salman Khan was working as an analyst for a hedge fund. If you had told him one of his projects would soon receive a $1.5 million investment from Bill Gates, he probably would have assumed you were talking about some creative new financial product.

It certainly would have been hard to imagine Bill Gates funding the primitive web videos that Khan was making in his free time to tutor his younger cousins. Less than a decade later, however, his Khan Academy has grown to encompass almost the entire curriculum from kindergarten through high school in thousands of short online videos. It is one of a handful of pioneering projects that are on the verge of revolutionizing general education.

When Khan began posting his math lessons on YouTube, he thought of them as supplements to the live tutoring sessions he gave to his cousins remotely over the internet. But as he recalled in a talk to the TED conference in 2011, "As soon as I put those first YouTube videos up ... a bunch of interesting things happened. The first was the feedback from my cousins. They told me that they preferred me on YouTube than in person."

At first Khan was perplexed as to why his cousins preferred the recorded videos to the live tutoring. Then he thought about it from the perspective of a struggling middle schooler: "Now they can pause and repeat their cousin, without feeling like they're wasting my time," he realized. "If they have to review something they should have learned a couple of weeks ago or maybe a couple of years ago, they don't have to be embarrassed and ask their cousin."

Something else surprised Khan about the YouTube videos he had started posting. Although he had made them for his cousins, the videos started getting lots and lots of views. People from all over the world left comments on Khan's postings telling him they were finally grasping concepts for the first time.

There's no doubt that the effectiveness of the videos is due in part to Khan's talent as a tutor and the personality that came through. His disembodied voice is reassuring; it sounds young and upbeat, like he's smiling as he teaches you math. He also has a knack for simplifying complex concepts—a quality you might not expect from a guy who holds three degrees from MIT and a Harvard MBA.

There is no flashy animation or fancy graphics in Khan's videos. While he speaks, he writes equations and sketches diagrams on a graphics tablet connected to his computer. Watching the lessons, you listen to Khan as his notes show up on your computer screen. At no point do you actually see him, so your brain spends no time analyzing the subtleties of a human face. Each video is short—about ten minutes—and much easier to digest than an hour-long lesson during the school day.

Khan soon realized these videos were reaching tens of thousands of people. "Here I was, an analyst at a hedge fund," he recounted. "It was very strange for me to do something of social value. But I was excited so I kept going, and a few other things started to dawn on me." First, he recognized, "this content will never go old.... If Isaac Newton had done YouTube videos on calculus, I wouldn't have to!" He was beginning to see how potentially disruptive the technology was in the hands of a good teacher. Millions of students could have access to the best tutor around.

Still, Khan thought of his videos as a supplement or a remedy for students who had missed something in the classroom. It had not occurred to him, he said, that teachers might adopt his work. To his surprise, however, he began hearing from them. "I started getting letters from teachers.... saying, 'We use your videos to flip the classroom. You've given the lectures, so now what we do is ... assign the lectures for homework, and what used to be homework, [we] now have the students doing in the classroom.'"

The "flipped" classroom let students learn at their own pace, repeating material they didn't understand or jumping ahead when they already knew where Khan was going. Students could then do practice problems ("homework") independently while at school and get help from teachers—or better yet, peers—when they got stuck.

By 2010, Khan had quit his job at the hedge fund and gone to work full-time on the videos, which had received tens of millions of views. Khan Academy added exercise modules alongside the videos to test students' understanding. Instead of testing once and moving on regardless of the results, as traditional schools have done, Khan Academy could ensure that each student mastered the important skills before trying to build on them. Once a student got ten questions in a row correct, the academy promoted them to the next lesson.

Khan realized there was a chance to implement this revolutionary model of "flipping the classroom" on a massive scale, making the former financier a co-teacher in thousands of schools.

Within a few years, Khan Academy approached the public schools of Los Altos, California, about integrating the content into their math classes. The schools agreed, "flipping" two fifth grade and two seventh grade classes.

This was a radically different approach to education. At Los Altos, students now worked independently through a long series of Khan Academy lessons as teachers kept tabs on their progress through a dashboard that showed where each child was in the curriculum. Khan Academy flagged students who were falling behind or stuck on particular lessons. Teachers could find out which problems were stumping students without having to ask them to their face what they didn't understand.

With such rich data about students' learning, teachers could help clear up targeted areas of confusion or even pair up peers to help each other. The progress they saw exceeded their expectations and challenged much of the traditional approach to education.

"In every classroom we've done, over and over again, if you go five days into it there's a group of kids who have raced ahead and a group of kids who are a little bit slower," Khan reported in his TED talk. "In a traditional model, if you do a snapshot assessment, you say, 'These are the gifted kids, these are the slow kids. Maybe they should be tracked differently. Maybe we should put them in different classes.'"

Students using Khan Academy to personalize their learning, however, broke this mold. "When you let every student work at their own pace," Khan said, "... over and over and over again, you see students who took a little bit of extra time on one concept or the other, but once they get through that concept, they just race ahead. So the same kids who you thought were slow six weeks ago you now would think are gifted."

Today, anyone can use the free dashboard features that the Los Altos classes developed with Khan Academy, and any parent, teacher, or tutor can sign up as a "coach." These tools could be a lifeline for millions of children who are trapped in schools that are failing them. The current system passes them through, whether or not they've learned the material, and then asks them to build on it the next school year. Khan Academy or pioneering platforms like it could end this cycle of failure. There's "no reason why it can't happen in every classroom in America tomorrow," Khan says.

Khan Academy doesn't just help rescue students who are falling behind. It also lets exceptional students speed ahead of the rigid, obsolete system that's holding them back. The Khan Academy website features a note from nine-year-old Harshal, who was bored in school even though he was already two grades ahead. "Now, I can always look forward to doing something that challenges me when I get home," he wrote. "Right now, I am doing derivatives in Khan Academy, which I never thought would happen."[2]

The website now offers more than four thousand videos on topics ranging from calculus to computer science to the Greek debt crisis to art history to American civics—almost all of them still recorded by Khan, though he's now supported by a sizeable staff behind the scenes. The motto blazoned on the homepage is "Learn almost anything for free." Six million students visit this virtual school every month.[3]

Khan himself speaks of the technology as a breakout not just for students in American schools but for people of all ages, all over the world.

Many visitors to Khan Academy are adults who are embarrassed that they need to brush up on skills they don't have but need in order to get a job. One such student wrote:

> I really can't begin to explain what seeing your site means to me. I'm a 36 year old man who 18 years ago wanted to be an engineer but as I progressed through school I got further and further behind.... As I watched your video explaining your vision I literally started crying because I saw that this is exactly what happened to me, the swiss cheese of the holes in my education added up to an insurmountable wall. I'm now on the way to becoming an engineer and I wanted to let you know that your videos have made a world of difference to me and mean that its possible for me to live a life I want rather than exist in a boring drudgery.[4]

Khan Academy is helping thousands of persons like this man achieve a better life.

It's also providing opportunities for people who live in places where they receive little if any formal education. Much of the content has already been translated into other major languages. Salman Khan imagines people on six continents tutoring each other and working together to learn new skills—"a global one-world classroom," he calls it. A classroom, maybe, with one amazing teacher and thousands of "coaches" to help.

That might sound ambitious for one guy. But if Khan has accomplished so much with the help of only a few friends and colleagues, we may have only begun to see the breakout in learning that he will lead.

Think about the best teacher you ever had. Maybe she was a math teacher who helped you see that you actually could do it. Or maybe she was an English teacher who knew how to interest you in books you'd never have wanted to touch. That teacher probably changed your life. She set you on a path to college. She shaped your interests and your future career. She changed the way you thought about the world.

Imagine if every child in America had access to that teacher or to someone like her. And not just in one subject for one year. Imagine if every child could learn from a teacher like that in every subject, every year of his or her education—if every child in America could learn from the best math, science, English, and history teachers there are.

It would be an education that no money can buy today. But soon it could be available for free to every child in America through a platform like Khan Academy. If it happens, it will be the beginning of a historic breakout.

* * * * *

Not everyone is eager to embrace the possibilities opened up by technology, and not everyone is pleased with Sal Khan and his Khan Academy. In 2011 and 2012, when Khan started getting lots of media attention for his work, including a place in *Time*'s list of the world's one hundred most influential people (with a profile written by Bill Gates), the old order's antennae perked up.

After all, if one guy with a tablet—a guy who professes no special pedagogical knowledge—can become "the world's teacher" and get millions of students to learn what they never learned before, what have our professional teachers been doing all these years? Perhaps even more threatening, what does it mean for the future of the profession if it only takes one (or a few) people to teach millions of students?

The prison guards of the past took notice and went on the attack.

In the summer of 2012, some teachers turned Khan's tool of choice, YouTube, into a weapon to use against him. Two math teachers appeared in a video critiquing one of Khan's lessons as it played on a screen in front of them.[5]

"Have you heard about this amazing new online teacher called 'Khan Academy?'" Teacher One says. "This is a guy who's putting all this content online, who's supposedly gonna *change education.*"

"Yes," Teacher Two replies, "my principal wants to do that."

"Well I figured we could take a look at one of these *videos,*" Teacher One continues. "I gotta imagine a guy who's got this much of a following has to be able to present this."

"Didn't Bill Gates give him millions of dollars?" Teacher Two asks. "This is gonna be amazing! Like, the best teaching millions of dollars can buy."

"Well Bill Gates calls him the best teacher he's ever seen, so I figure we have a lot that we can learn from this," Teacher One says sarcastically. "Let me get it started."

Salman Khan's voice begins, "Welcome to the presentation on multiplying and dividing negative numbers...."

"Only eight and a half minutes! I usually teach for hours on this," Teacher Two cracks.

The pair then continue through Khan's whole video (one plucked from several thousand on the website) and criticize things like Khan's handwriting (one of his arrows is too curved for their taste—it looks like a letter "d"), his choice of numbers ("1 x 1" might be taken for "$|x|$," the absolute value of x, they say), and the order in which he explained his examples.

At one point, Teacher One laughs at the idea that his students still "would be paying attention at this point."

The video went viral, racking up tens of thousands of views, generating a Twitter hashtag, and sparking a competition for other teachers to produce similar video attacks on Khan Academy.

It was written up in the *Chronicle of Higher Education*, Slate, *Wired*, and the Huffington Post.[6] *Education Week* published a story headlined "Don't Use Khan Academy Without Watching This First."[7]

Khan Academy replaced its video on multiplying and dividing negative numbers with one that addressed some of the complaints, but the criticism that emerged in the subsequent press coverage was more fundamental. Khan, it was said, is all procedures; it does not address deeper concepts in mathematics.

The *Washington Post* published the most hysterical—and probably the most widely distributed—anti-Khan rant a month later in an article by Karim Kai Ani, "a former middle school teacher and math coach."[8]

Kai Ani wrote that the big problem with Khan Academy is that "the videos aren't very good." They are simplistic, he said. The way they teach the concept of *slope* as "rise over run" is terribly damaging, he suggested, because slope isn't "rise over run" at all—that's merely a way to "calculate slope." (Never mind that teachers have taught "rise over run" for generations—largely because, as Khan points out in his response, "Slope actually is defined as change in y over change in x [or rise over run].")[9]

It is this kind of sloppiness, Kai Ani argued, that provokes "experienced educators ... to push back against what they see as fundamental problems with Khan's approach to teaching." The real issue with Khan Academy isn't the disruptive technology after all. It is Khan.

These experienced educators are "concerned that he's a bad teacher who people think is great." They are "concerned that when bad teaching happens in the classroom, it's a crisis; but that when it happens on YouTube, it's a 'revolution.'" The truth is, Kai Ani concluded, "there's nothing revolutionary about Khan Academy at all."

Kai Ani and his fellow teachers have a point that the process of solving math problems should be supplemented by conceptual explanation. They might even be right that the cherry-picked video on multiplying and dividing negative numbers was heavy on the process. But that criticism doesn't fit the work of the Khan Academy as a whole, which includes lessons on cryptography, Baroque art, and black holes.

More to the point, who are these "experienced educators" to call Khan a "bad teacher who people think is great"? Millions of people are voting with their feet because they can tell that they're learning from Khan what they haven't learned in school.

It's a sign of everything that is wrong in American education that the representatives of our failed educational system feel entitled to decide whose teaching meets guild standards while ignoring the actual results with students.

The carping of "experienced educators" is especially hard to stomach in light of the mediocrity (or worse) that prevails in many American classrooms. Even if Sal Khan's handwriting isn't perfect or his examples aren't always bulletproof, isn't his academy miles ahead of the middling or failing schools that are leaving so many young people intellectually crippled?

One could easily get the impression that the detractors aren't hostile because they think Khan is a bad teacher at all, but because they recognize he's a great tutor doing much of their job for free.

Of course, not all teachers have spurned Sal Khan. The schools that have adopted Khan Academy to "flip" their classrooms are a major part of his story.

But the prison guards' assault on Khan Academy in YouTube videos, education journals, and the *Washington Post* is not just funny satire.

It is pernicious. The prison guards are trying to kill the future. We know they're trying to kill it, because they're not saying, "Gee, what an amazing start for one guy trying to cover the whole curriculum! Let's make it better together, and then we will really have something."

No, they're saying, "You can't trust this! He isn't certified. And you don't understand that you're not *really* learning from him. You need teachers in classrooms teaching just the way teachers have always taught. We will not have any of that new stuff here."

The prison guards don't want a better version of Khan Academy. And they don't want their colleagues "flipping" the classroom and rocking the boat. They want to stop any change—end of story.

Not all teachers are prison guards of the past. Thousands of teachers in schools across the country are trying every day to find better ways to do their jobs. But even the most innovative teachers are trapped in an unchanging system that passes along to them unprepared students and expects them to work magic.

Still, as the reaction to the Khan Academy reveals, many of America's schools are in fact full of prison guards, jealously protecting their own privileges.

The teachers' unions in particular are determined to impose the failing present on America's students. They exercise their considerable power to block charter schools, school choice, and other reforms that might give millions of students greater opportunity. They impudently resist competition, accountability, or quality assurance in the education of our children.

Too many teachers protect themselves at the expense of their students, giving up on their mission and turning themselves into prison guards. How else can you explain forcing students to endure a bad teacher for years rather than allowing the teacher to be fired? Or chaining kids to schools that are failing rather than letting them try someplace where they might have a chance to succeed? How else could they oppose paying great teachers more than poor teachers? Or teacher evaluations that reveal whose students are learning and whose are not?

Virtual Charter Schools

Technology has done surprisingly little to change education. The education establishment, by and large, is interested in using technology to augment the existing system—thirty-five kids in a cinderblock classroom—rather than to transform it. Some schools have adopted digital whiteboards and installed video projectors, but they have obstructed the fundamental change that technology makes possible. Writing on a fancy digital screen at the front of a classroom instead of a chalkboard does not solve the problem that half the class is bored and the other half is lost.

A radical innovation that takes up where Salman Khan leaves off is the virtual charter school movement. Like brick-and-mortar charter schools, virtual charter schools are an alternative to neighborhood public schools, but that's where the similarity ends.

The students in a virtual charter school like Florida's Virtual Academies or Pennsylvania's Agora Cyber Charter School don't climb into a big yellow bus each morning. They head to the family room and log onto their computers. They make their way through videos on dividing with decimals, complete exercises on calculating area, and perform virtual chemistry experiments—all at their own speed, moving quickly through what comes easily and taking more time with what's difficult. They can access the material at any time of day, so they enjoy a flexible schedule and pace.

Work in schools like Agora or FLVA isn't all independent, however. Students meet online at set times for classes with real teachers (also working from home) and other pupils. Communicating through microphone headsets, they can attend lectures and participate in discussions.

Students often report that they receive more personal attention from teachers in their virtual charter schools than in their previous schools. The instructors can "drop in" electronically at any time and clarify concepts one-on-one. When they suspect a student does not fully grasp a topic, they call to quiz him a little.

Although virtual charter schools replicate some of the most important features of traditional schools, they can look more like homeschooling than classroom education. Students benefit from their parents' involvement in the school day. Mom or dad is a "learning coach" who makes sure they're not playing Angry Birds when they should be discussing Atticus Finch. The coaches are in continuous contact with teachers to monitor their children's progress. There's no hiding poor test scores, inattention, or misbehavior from mom and dad.

With this careful guidance from teachers and parents, students in virtual charter schools are free to pursue a radically different education. Students in traditional schools rarely have time for an apprenticeship at a hospital or radio station, but with a flexible schedule and the extra hours a virtual school opens up, these opportunities become real possibilities. It's easy to imagine groups of online students meeting face-to-face several days a week to build a robot or visiting a different local museum every Wednesday. Others might form teams to publish a magazine or to plant a garden. There is a world of learning opportunities outside the traditional classroom.

More than a hundred thousand[10] students nationwide are already attending virtual charter schools like FLVA and Agora. Most of them are powered by software from K12, an online education company that produces thousands of hours of high-quality content in the K–12 curriculum.

Critics of schools like Agora cite below-average test scores and high dropout rates. Brandishing a report that more than half the students at Agora are below grade level in reading and math, they condemn the online model. It's true that such numbers raise concerns, but the criticism seems shortsighted. Today virtual charter schools attract primarily those students whom the conventional public schools have failed. Many have switched to virtual charter schools after spending years, perhaps even their entire education, in deplorable schools. It shouldn't come as a surprise that large numbers of such students are below grade level.

This is not to say virtual schools are without any problems. Some teachers have complained about having to handle too many students.

Others worry that too many kids are passed through without really understanding the material. No doubt there is lots of room for improvement. The important question is not what virtual charter schools are today, but what they could become tomorrow. The technology won't always be limited to students who have a parent at home. Someday teachers themselves could look more like "learning coaches."

The possibilities are exciting. The schools are only beginning to integrate the continuous assessment Salman Khan spoke about, ensuring that every student understands crucial concepts before trying to build on them. Combined with the outside activities that a flexible schedule permits, the individualized structure and pace of the curriculum should make the virtual charter school education uniquely engaging. The almost infinite entertainment options that compete with school for kids' attention make this a considerable advantage.

Many children these days routinely play games on iPhones by the time they learn to talk. Before they are in kindergarten, they are used to getting the information they want when they want it through a computer or a tablet that talks to them. Three-year-olds will know how to ask Siri to read them *Curious George*. Long textbooks and dense fifty-minute lectures are going to seem intolerable to those kids by the time they reach sixth grade.

Something will have to change. Computer-adaptive algebra games and an online discussion of *The Adventures of Huckleberry Finn* at home in the morning followed by an afternoon of building drones with other students sounds like a pretty exciting breakout in learning—and one that has a prayer of holding kids' attention. It will be even more game-changing if it takes students out of failing schools and matches them with teachers and classmates who can help them succeed.

Like good prison guards, the teachers' unions are trying to stop this transformation from ever getting off the ground. In state after state, companies like K12 have had to fight costly battles to have the chance to compete. The prison guards didn't try to make Sal Khan's math lessons better, and they aren't trying to figure out how we can improve virtual schools. They're trying to kill them, depriving millions of students of a potentially better future for the sake of their own privileges. Whatever challenges virtual charter schools face, students in thousands of traditional schools are performing far below grade level and eventually giving up. Yet the unions are suing to keep those schools *open*. One in four students nationwide does not graduate high school, and in some places the rate is much higher.[11] Nearly 40 percent of students in Chicago drop out of school; among African Americans, that number is over 60 percent.[12] One out of three fourth graders cannot read, scoring "below basic" on literacy tests. One-third of eighth graders and 38 percent of twelfth graders read below grade level.[13] The unions have nothing to boast of, and the prisoners of their schools could hardly do worse in a virtual school.

We haven't begun to see the real breakout that virtual schools could offer. Their potential to outperform traditional schools is impressive, but the breakout will come only when they integrate the next big leap in education: learning science.

Bror Saxberg and Big Data

If you glanced at Bror Saxberg's résumé, you'd probably assume that he's a cloistered professor at an elite university. He holds two bachelor's degrees from the University of Washington, one in mathematics and one in electrical engineering. As a Rhodes scholar, he received a degree in mathematics from Oxford, and he washed it all down with a PhD in electrical engineering and computer science from MIT, which he earned simultaneously with an MD from Harvard Medical School.

Despite his twenty-four-karat academic credentials, Saxberg is no snob about education. Indeed, he has chosen a career many of his peers from Harvard, MIT, and Oxford might (wrongly) view with disdain. The chief learning officer for Kaplan, one of the largest for-profit education companies, Saxberg is responsible for improving the learning of more than 1.5 million[14] students worldwide. I got to know him while working with Kaplan on a series of short courses for NewtUniversity.com.

Google can determine traffic conditions on thousands of roadways by analyzing where millions of Android phones are. The Obama campaign optimized its donation form by live-testing hundreds of variations. Saxberg thinks that "big data"—generated by millions of students online—can likewise transform education.

With the zeal of an entrepreneur, Saxberg talks about presenting every concept to every student in the way that is best suited for *that* student. He recently tested Kaplan's approach to preparing students for the logical reasoning problems on the LSAT, which many find particularly tricky. The company had invested heavily in an hour-long video that was loaded with fancy animations to illustrate the problems, but Kaplan's "learning engineers" (people trained in learning science but applying it to practical problems at scale) thought there was room for improvement.

Using Amazon's "Mechanical Turk" platform,[15] the Kaplan team created a task that directed part of his test group to the hour-long video and then gave them a few of the LSAT questions. The second part of the test group read through a couple of "worked examples"—annotated text versions of how someone solved the problems correctly, shown by Australian learning researcher John Sweller to improve problem-solving—instead of watching the video. A third group got no explanation at all.

For a few thousand dollars, by the end of a day or two of testing, Kaplan had a randomized, controlled study comparing its video approach with a new alternative—approximating research that traditionally would take years and, in traditional university settings, cost tens of thousands of dollars or more. Saxberg's team quickly discovered that there was little difference between the performance of the students who watched the video and the performance of the students who had no preparation. The video was ineffective.

The subjects who read through the worked examples, however, performed considerably better, and their preparation took only about fifteen minutes, compared with a full hour of video instruction. Kaplan is adjusting its method for teaching the LSAT problems accordingly.

Saxberg's results reflected what he says science shows about how we learn: the most important factor is usually not the teacher or even the student's innate talent. "What really makes the difference," he says, "is deliberate practice." By and large, people master the multiplication table or factoring quadratic functions the same way they become great basketball players or excellent airplane pilots: many (not all) of the key decisions and tasks need to be practiced and practiced, with feedback, until they are hardwired into their brains. An hour spent watching a video makes little difference, but an extra thirty minutes of practice problems helps dramatically.

Although this discovery, in itself, has nothing to do with technology, it points us to a more technology-based education system, Saxberg argues. An online environment, he told me, enables a "competency-based model of instructional learning rather than a schedule-based model, one based on seat-time." In such a system, students advance not when the class schedule dictates, but when they have actually learned the material—and they get sufficient practice and feedback, made efficient by technology, for each of them to get there. The same technology, Saxberg said, "enables things like … data-gathering modules and diagnostics that tie into banks of hopefully usefully designed training materials in ways that just weren't possible before, when you just had to work with an individual instructor who was in your face because there was no other way to get the learning that you needed."

Like Salman Khan, Saxberg sees the explosive potential for personalized learning. "You can start to track how people are doing and you can speed up or slow down the pace at which you're giving them examples. There is a lot of great stuff to explore that hasn't even begun to be explored."

The sea change happens when you plug tens of thousands of students into such a system, he said. The avalanche of performance data enables those "learning engineers" to observe and test patterns of what is working and what isn't and for what kinds of students. Which students should have the video feed turned off? Who works best in a small group, and who learns best on her own? Who needs videos or practice types Q, R, and S as opposed to A, B, and C? What happens if you give both Q and B—does performance go up, and is it worth the extra time? For whom?

In a large population—say, a big state education system—educators can quickly gather evidence about the optimal ways to teach their material. "Having a lot of learners that are in a systematic learning environment gives you the potential to do a lot of work," Saxberg said. "Right now at Kaplan, we are setting up to do … maybe north of one hundred randomized controlled trials each year. We have some courses that have more than a thousand students starting every month. So you think about that: a thousand students starting every month, you should be able to do ten hundred-student randomized controlled trials every single month, so even a single course might get you a pace north of one hundred controlled trials in a year."

Today, little classroom teaching is based on solid evidence of how students learn most effectively. Computerized learning environments tied in with teaching will finally let educators systematically apply the insights of science to the classroom. "We are applying results that have been around for decades in the world of learning science research," Saxberg says, "about media and images and texts and the structure of the pieces and how you do some of the feedback and worked examples. Good evidence from laboratories suggests that these things make significant differences to learning, but they are not known or used at scale." Soon, he says, parents could demand to see the evidence that supports how their children are being taught.

In an online learning environment, you could apply new insights to the entire system, hundreds of thousands of students, just as fast as you discovered what works for various types of learners. If one teacher had an idea for teaching fractions that improves practice results by 5 percent in certain kinds of students, all of them could benefit from it the next week.

We could do all of this today. There is no good reason that every child in America can't receive the effective education that the analysis of big data makes possible. But how will teachers' unions react to the data-drenched reforms that Bror Saxberg describes? Judging by their responses to every other proposed reform—teacher evaluations, merit pay, and the abolition of tenure, to name a few—we can expect a fight to the death. The victims, of course, are the students, who will learn less than they need to compete in the modern world.

Saxberg isn't pointing his fire hose of data at K–12 education, however. Kaplan, after all, runs the world's second-largest online university. Despite the scorn with which many professors at the "elite" universities view for-profit colleges, Kaplan is in some ways far ahead of them. It has found a way to give people the education and skills they need without breaking the bank. As the pace of innovation will require a workforce of lifelong learners, this is essential. Just as Netflix allows viewers to log on and view what they want on their own schedule, online education enables people with full-time jobs to learn what they need on their own schedule.

Compare online students with those at traditional universities, who take on tens of thousands of dollars of debt to finance their education. What do they get at Harvard or Berkeley for that kind of money? They probably get a professor who views teaching introductory biology as nothing short of purgatory. He is likely to be far more interested in his research into gametogenesis or in his forthcoming appearance on PBS than in teaching college freshmen the basics of plant photosynthesis. There's a good chance, then, that he will recycle the same PowerPoint he has been using for the last twenty years, reciting the same lecture he has always given, without much thought about whether it's the "optimal" way to teach the material or not. Even if a few professors are dedicated to the success of their students, the research-based tenure system means they have to regard it as a hobby, not their primary job. In the clearest signal of its bias against teaching, the system rewards the "best" faculty by removing teaching from their workload.

There is a deeper problem as well: "Historically, we conflated being an expert with being an expert on how to teach," Saxberg told me. "The two are totally different. Experts have been doing their work for decades, often, and 70 percent or more of what they do is done in automated ways by their minds." Many of them "tend to only teach what they think consciously about, which means they are going to leave out 70 percent of what a novice actually needs to know," he said. "So this is one of the problems with the current education system, where experts teach the class, and the Nobel Prize winner has no idea how to describe what he knows so deeply, it's become non-conscious. Nobody quite understands what he just said, nor realizes what has been left out."

If students are lucky, their professor might hold office hours twice a week. The bolder students can show up to annoy him if they didn't understand what he said in class, in which case he will patronizingly direct them to the nearest teaching assistant. If that doesn't help, they've got the textbook. (No wonder traffic is so heavy at Khan Academy.)

In contrast to the challenged state of traditional higher education, the nuts-and-bolts approach Bror Saxberg describes is downright refreshing.

Unlike Kaplan, America's elite universities usually have no one whose job is to make sure professors are teaching their courses in the most effective manner. They don't try to follow their students from course to course and then after college to see if what they learned is helping them to succeed. Where are the "learning engineers"?

Even if Columbia or Cornell did have someone to compare the test scores of Professor Watson's section of Calculus 101 with those of Professor Price's section, what could that person do? If he told Professor Price, "You need to do it the way Professor Watson is doing it, since his students consistently score 30 percent higher than yours," Professor Price would probably tell him to get lost. Everyone in the modern world of higher education understands that lecturing the debt-laden student body is what the faculty does to pay the bills but that it's a secondary concern at best.

Udacity

Why are Harvard, Princeton, and Yale behind Kaplan and the University of Phoenix when it comes to personalizing their courses or making their lectures available to students at any time of day? Why have they largely failed to adopt competency-based adaptive learning systems so students can set their own pace?

Until recently, the accreditation guild that has fueled much of the student loan bubble protected America's leading universities from meaningful competition. The average cost of a college education has increased twelvefold, or 1,120 percent, since 1978, according to Bloomberg—four times the rate of inflation.[16] In 2013, two-thirds of graduating college students had outstanding school loans, with an average balance amounting to 60 percent of their annual salary.[17]

For-profit education companies (which are constantly under regulatory assault by the old order) are often accused of churning students through the system and giving them a degree without teaching them much, leaving them burdened with debt they will struggle to pay off. It is often more accurate, however, to level this accusation at traditional "non-profit" universities, including America's leading institutions of higher education, than at their for-profit competitors. As the work of

Salman Khan and Bror Saxberg suggests, this unsustainable model could be on the verge of collapse.

One pioneer working to bring down the cost and improve the quality of higher education is Sebastian Thrun, a vice president at Google and a professor at Stanford. He already has a few world-changing inventions under his belt. He helped create the self-driving cars that have safely navigated six hundred thousand miles of California roads—a breakthrough that I explore in chapter six—and he helped pioneer Google's Street View technology, which has indexed images of almost every inch of public road in the United States and many countries around the world.

What could prove to be Thrun's most disruptive project, however, began not at Google but at Stanford. Along with Google's director of research, Peter Norvig, he taught Stanford's computer science course on artificial intelligence. The pair decided to offer a version of their course online for free, unsure of how many people would sign up.

Before they knew it, more than 160,000 people from all over the world had registered.

As Thrun recounted to the *Wall Street Journal* last year, there was just one problem: "I had forgotten to tell Stanford about it.... Stanford said 'If you give the same exams and the same certificate of completion [as Stanford does], then you are really messing with what certificates really are. People are going to go out with the certificates and ask for admission [at the university] and how do we even know who they really are?' And I said: I. Don't. Care."[18]

By the end of the semester, twenty-three thousand of the original 160,000 had completed the full course. The best Stanford student was ranked 411.[19] The results were stunning to everyone. As Thrun explained it to me, "The fact that we brought twenty-three thousand people to the finish line meant that in this specific quota I taught more people artificial intelligence than all the other professors in the world combined." Reaction to this remarkable achievement was mixed, however. "To the world, it mostly meant, 'Wow, we're entering a new era of education where things become accessible, affordable, and also transparent, and as a result higher quality,'" Thrun said. "To some, it means, 'Wow, did I just lose my job?'"

Thrun was amazed by the results of his experiment, which became one of the first so-called MOOCs (short for "massively open online courses"). He was so excited by the new possibilities that he cut his work at Google to part-time and founded an education company, Udacity, with the goal of reducing the cost of a college degree by 90 percent. "I think we found the magic formula," he told me. "That may sound pretentious, but the magic formula for me is high-quality education at scale ... I think we owe students a personalized, high-quality experience."

In addition to a version of Thrun and Norvig's original class on artificial intelligence, Udacity now offers college-level introductions to computer science, statistics, physics, and psychology, as well as a smattering of more advanced courses. In "Differential Equations in Action," students discover how to use mathematics to "rescue the Apollo 13 astronauts, stop the spread of epidemics, and fight forest fires." In another class, students develop sophisticated browser-based computer games.

The most-exciting courses on Udacity are those offered in conjunction with traditional, credit-granting universities. Udacity took a radical step in 2013 when it partnered with San Jose State University in California to offer introductory-level math, computer science, and psychology courses for credit, enabling San Jose students to apply their work toward a real degree. The cost? One hundred fifty dollars for each course.

It was a comment by Bill Gates that motivated Thrun to work with San Jose State. The early Udacity model, Gates said, worked only for extremely self-motivated people, maybe 5 to 10 percent of the learning population, since they had to work independently. "I took this to heart and thought about this," Thrun said. The San Jose project was his first effort at solving the problem.

"We picked exactly the opposite" of the top 10 percent Bill Gates had in mind, "those students that traditionally are the hardest to reach and educate," Thrun told me. "We picked inner-city high school students from low-income neighborhoods, and they come with enormous ... handicaps. We deliberately even picked a group of students that not only had failed the entrance exam for college in mathematics, but also had failed the subsequent remedial education that they had to undergo to make up for the remedial math. Out of those kids, about 6 to 10 percent can stay in college. So these are basically kids on the way out from college. And we picked those because we wanted to understand how to make education work for everybody."

The Udacity students at San Jose had an 83-percent finishing rate—including students who had been on the verge of dropping out of college—as opposed to the 5-percent rate in an ordinary MOOC. In each class, unfortunately, no more than 44 percent of students managed to pass the final exam. But considering the pool from which Udacity intentionally drew, this figure suggests possibility: in some courses almost half of the students who were "on their way out" actually passed the class.

Of course, the media didn't see it that way. I talked with Thrun just days after those results were released and critical stories from all the usual suspects were circulating around the internet. According to the *Chronicle of Higher Education*, the numbers had been "leaked ... to *Inside Higher Ed* and *The Chronicle*" by "the California Faculty Association, a union that has been vocal in its criticism of the Udacity partnership."[20]

Thrun, however, was upbeat. He had expected they would learn a lot by taking up Bill Gates's challenge, and they had. In short, they realized that "to be good in math requires more time than is normally given at college." Students who were already behind needed more flexibility and more time to catch up—which online education was uniquely able to offer—but it was time they didn't have in this case. They also found that students loved the online model, supplemented by people "on the ground," and preferred it to traditional classes. That system "was able to really get these kids' attention and really drive them through the program all the way to the end," Thrun said.

That same year, Udacity announced a program aimed at a very different group—a partnership with Georgia Tech to offer a master's degree in computer science for $7,000, beginning to end. That's a graduate degree from one of the top-ranked computer science programs in the country, and it includes online course content supplemented by a lot of real people who have student contact and help them work through problems.

That kind of competition from Udacity and other online-learning options should put pressure on the higher education establishment, which charges students in the neighborhood of $200,000 for a four-year degree. If Udacity offered enough courses at $150 apiece to constitute a degree, it would cost well under $10,000. An education at that price would be a game-changing opportunity for millions of people who today can't afford college or who struggle to cope with the debt. It would also have enormous implications for affordable, convenient learning for adults.

In 2013, President Obama announced a government-run, centralized system for rating institutions of higher education that will tie federal funds to a bureaucratic measurement of a university's "value." This is a step in the wrong direction. We should encourage more competition from upstarts like Udacity rather than locking the education system into a government formula.

Asked if he was prepared for the resistance of his Stanford colleagues and the other gatekeepers of higher education, Thrun replied, "Someone's got to do it. I have nothing to lose. We can help get kids access to good education, important education for the nation. People understand America is the country of innovation. We've always innovated our way out of crisis. We have to innovate. We can't be driven by people who are skeptics."

Thrun is optimistic that projects like Udacity can overcome the opposition of the prison guards in education and offer American students—and for that matter, students all over the world—a real chance to break out. "Online education will way exceed the best education today. And cheaper," he told the *Journal*. "If this works, we can rapidly accelerate the progress of society and the world. If you think Facebook is neat, wait five to ten years. So many open problems will be solved."

CHAPTER THREE

BREAKOUT IN HEALTH

How much would you pay to have a heart attack? How does $45,000 sound—$7,000 out of your pocket and about $38,000 out of your insurer's? Or diabetes? The average case will run you $9,975 every year. You can have asthma for just $3,300 annually. Other common health emergencies are *really* expensive: the lifetime cost of surviving a stroke is upward of a hundred thousand dollars. The average case of breast cancer is $128,500.[1]

If you don't face these health problems, there's a good chance you have friends or family members who do. Not one of them was looking to spend his hard-earned money, not to mention years of his life, dealing with an expensive medical problem.

You may be amazed to learn that these prices are not inevitable or immutable. They are a product of a badly functioning system that has gotten away with dramatic overpricing in the absence of competitive markets. The very regulatory system health professionals complain about has in fact been hiding the enormous inefficiencies that make American healthcare unnecessarily expensive.

Dr. Devi Prasad Shetty, a leading heart surgeon in India, charges $1,583 for coronary bypass surgery. The distortions in the American medical system produce a price of $106,000 for the same procedure at the Cleveland Clinic.[2] Assume for a moment that the Cleveland Clinic is better. Is it *sixty times* better?

Dr. Shetty's achievement is not isolated or accidental. His procedure cost twice as much twenty years ago. Then he set about methodically to eliminate waste and to lower costs so that more Indians could afford it. He now has twenty-one hospitals delivering this kind of care, and he hopes to bring the price down to eight hundred dollars in the next decade.[3] There is every reason to expect that the Cleveland Clinic's price will be even higher by then.

Despite America's failure over the last four decades to control healthcare costs, however, there is a chance that our medical system will offer better quality and lower costs in the years to come.

In the past few years, the medical world has been rocked by two starkly different approaches to improving patients' health. You've almost certainly heard of the first one, Obamacare, which Americans were promised would help them pay those pricey tabs. This approach comes with lots of bureaucrats in Washington, more regulations, IRS agents to wade through your medical bills, and boards of experts to tell you which treatments you may and may not have. Under Obamacare, the Department of Health and Human Services, the Food and Drug Administration, Medicare, Medicaid, and the rest of the healthcare bureaucracy will take over your doctor's office. You might get a subsidy to help pay for that forty-five-thousand-dollar heart attack (then again, you might not), but you and your doctor will certainly make fewer choices about your care, while bureaucrats will make more. And by necessity, a large bureaucracy must make sweeping decisions for whole populations at a time.

While you have heard plenty about Obamacare, you might not have heard as much about a second approach to reforming our dysfunctional healthcare system. That's because it sneaked up on us. While we were debating about insurance coverage and Medicare bills, many of the transformational breakthroughs in medical science that we had long been promised—personalized medicine, treatments based on genomics, and regenerative therapies—approached the stage of development at which they could be offered to average people. These technologies have the potential to give us much longer, healthier lives in a very different world—a world in which diabetes, heart attacks, and even cancer are either completely avoidable or merely short-term inconveniences.

Obamacare's hyper-bureaucratized and depersonalized approach to healthcare and the new technologies of personalized medicine are headed for a painful collision in the near future. As medicine moves toward treatments that are exactly right for *your* problems with *your* body, our system for regulating and paying for healthcare is moving in the opposite direction—taking decisions away from *you* and *your* doctor and giving them to "experts" you have never met.

Many of the breakthroughs you will read about in this chapter must clear some technological hurdles before they are ready to transform medicine. When they finally come to fruition, they may look different from what we're expecting. Other revolutionary technologies are here now, waiting for us to adopt them. What's clear is that within the next ten years, we could achieve a breakout in health that would have been almost unimaginable even one generation ago. In this new world, the prison guards keeping us trapped in a bureaucratic big-government healthcare system may be the greatest public health threat of all.

Personalized Medicine

If there's a Moses of medicine, it might be Dr. Eric Topol. During a long tenure as head of the cardiology department at the Cleveland Clinic, Topol led clinical trials for a number of the major drugs on the market today. He was the first person to identify genes related to heart disease, and he is one of the ten most-cited researchers in medicine, according to the Institute for Scientific Information. Topol might be best known, however, for being the first to question the risks associated with the blockbuster drug Vioxx, which was eventually pulled from the market.

Today, Topol is the director of the Scripps Translational Science Institute and a leading thinker about personalized medicine and the future of his field. At the beginning of his book *The Creative Destruction of Medicine*, he describes what stands between us and a medical promised land: "Beyond the reluctance and resistance of physicians to change, the life science industry (companies that develop and commercialize drugs, devices, or diagnostic tests) and government regulatory agencies are in a near paralyzed state, unable to *break out* of a broken model of how their products are developed or commercially approved. We need a *jailbreak*."[4]

We can achieve that breakout, Topol writes, by using the last decade's giant strides in computer and information technology to rethink fundamentally how we do medicine. So far, he argues, "we've done virtually nothing to embrace or leverage the phenomenal progress of the digital era." But the technology will soon enable something revolutionary: "For the first time in history," he writes, "we can digitize humans."[5]

If that language conjures some frightening images of a brave new world, don't worry. Topol isn't talking about downloading your brain into the Matrix. But some of his ideas sound almost as fantastic. Within a matter of years, he says, every patient in America could browse his own complete genome on an iPad. For more real-time data, tiny sensors on the skin—or even nanosensors in the bloodstream—could "remotely and continuously monitor each heart beat, moment-to-moment blood pressure readings, the rate and depth of breathing, body temperature, oxygen concentration in the blood, glucose, brain waves, activity, mood—all the things that make us tick." These data will be available not just for patients in the hospital, hooked up to cumbersome monitoring contraptions, but for everyone at all times, accessible constantly with apps on our smartphones.[6]

These breakthroughs aren't decades away; many of the technologies are here now. An NBC News report in January 2013 showed Topol as he popped a special case onto the back of an iPhone—a case much like the ones many of us use to protect our devices—and handed it to a patient. Within seconds, the man's fingers touched a pair of metal contacts, and a complete, bouncing cardiogram appeared onscreen. An indicator in the corner counted sixty-six beats per minute. The device, according to the report, cost only $199—the price of just two traditional echocardiograms with a technician.

"I'm used to having an ECG machine hooked up to me, and shaving my chest and putting stickem on there and putting on the electrodes or whatever," his patient commented. "This is incredible."

A moment later, Topol used a handheld ultrasound device to examine the man's heart, a procedure he said would normally take place separately and cost eight hundred dollars. Later, Topol ate a handful of tortilla chips and watched on his phone as his blood glucose level responded, with data transmitted from a postage stamp–sized sensor taped to his abdomen. Any of these data and more, he said, could be monitored by doctors remotely: "You can do blood tests, saliva tests, urine tests, sweat tests—all kinds of things—through your phone."[7]

Doctors will have access to all of this information, Topol predicts, plus our whole medical history in one file detailing who we are, biologically. They'll use it to personalize treatment, monitoring patients who are genetically or constitutionally predisposed to certain problems and delivering custom drug cocktails or targeted treatments as needed. Nanosensors in the blood, for instance, might alert doctors to an impending heart attack. They could respond with precise doses of blood-thinning medication. Other sensors might monitor for early signs of breast cancer in the bloodstreams of patients who are especially at risk, while low-risk women could avoid the frequent screening procedures.

An even more revolutionary change, Topol says, could be that medicine won't begin and end with a doctor anymore. *We* could have our own data. And we could harness it in entirely new ways, such as feeding it into apps that will diagnose us automatically—analyzing our ECG for early warning of a heart attack, for example, or applying pattern recognition to an image of our throats, confirming that we indeed have tonsillitis. In harder cases, we might try to crowdsource a diagnosis (perhaps by uploading segments of our personal health data to an internet forum) before we go to a doctor.

The digital era is making the old gatekeepers of information obsolete in almost every field of knowledge, and health information cannot remain an exception forever. It's not hard to imagine that one day a

"Wikipedia for medicine," using sensor data and genetic profiles uploaded by tens of millions of people, could power a free, computerized version of Dr. House—far exceeding the capacities of even the best diagnosticians today. Such a system could keep much more granular data on treatments and results than is now reflected in clinical trials, allowing your smartphone to give you a more precise prescription than any doctor currently could.

Such intense personalization is the opposite of how doctors practice medicine today, which Topol describes as "maximally imprecise." In what he calls "imprecise medicine," doctors take their cues from guidelines issued by insurance companies and government agencies, guidelines that by definition are "indexed to a population rather than an individual."[8] Genomics and the new sensors (what Topol calls "wireless medicine") are pushing medicine in a completely different direction.

Topol points to Lipitor, one of the most commonly prescribed drugs in the world, as a case study of imprecise medicine's shortcomings. Advertisements for the drug proclaimed a 36-percent reduction in the risk of heart attack—an apparent triumph over a leading cause of death in the United States. The fine print, however, tells a different story: "'[I]n a large clinical study, 3 percent of patients taking a sugar pill or placebo had a heart attack compared to 2 percent of patients taking Lipitor,'" he writes. In other words, for "every one hundred patients taking Lipitor to prevent a heart attack, one patient was helped, and ninety-nine were not."[9]

In the days before genomics, wireless medicine, and "big data"—the days when population medicine was the best we could do—it might have made sense to prescribe a drug to a hundred "at-risk" persons on the chance that one of them would be helped. If your doctor warned you that you were at risk for a heart attack, you would want the latest and greatest drug available. And you wouldn't be satisfied if your doctor tried to reassure you, "But chances are you'll be *fine*." When you're told you could have a heart attack, you want a prescription.

The medical industry today relies on this type of treatment, prescribing drugs to millions of persons who meet some broad criterion, knowing that a subset of the group will benefit. For the most part, pharmaceutical companies produce drugs that can be dispensed to these large populations.

The FDA's outdated clinical trial standards all but require it. And most doctors consider randomized controlled studies on large populations to be the paragon of evidence-based medicine. After all, they produce genuine statistical proof that at least *some* patients will be helped, maybe even saved, by a drug—and that is nothing to sneeze at.

But by now we should be doing much better. A decade's worth of giant strides in science and technology should have made population medicine obsolete. Physicians should be able to identify (using sensors in the bloodstream, for instance) which three of a hundred "at-risk" patients are in fact approaching a heart attack. The other ninety-seven, whom we consider "at risk" but who will never actually suffer a heart attack, should be unhappy about taking a costly prescription that may also have nasty side effects. For those patients, the drug is no good at all. "What constitutes evidence-based medicine today," Topol concludes, "is what is good for a large population, not for any particular individual."[10]

A modern medical system would place a high premium on producing a complete genetic profile for every patient and on integrating as rapidly as possible the personalization, wireless medicine, and cost-saving devices that Topol demonstrated on television more than a year ago. Every patient who is currently taking thousands of dollars' worth of pharmaceuticals each year would become a walking, breathing, data-gathering medical machine (data that patients, of course, would own). Each person would get the most targeted, customized treatments available. Healthcare, like education, should be moving quickly toward mass personalization.

* * * * *

But since the passage of the Affordable Care Act—Obamacare—our medical system has been heading in the opposite direction. Just as individualized medicine has emerged as a practical possibility, the prison guards in Congress and in the vast federal bureaucracy have seized control of *more* of our healthcare. The so-called "reform" of healthcare has frozen an already problematic system in the past—and made it worse.

A centralized bureaucratic healthcare system was a bad idea even in the era of "one-pill-fits-all" medicine. It consistently resulted in lower-quality healthcare at a higher price. But such a system is positively deadly in an era when we can customize treatment to each person's age, height, weight, and gene type, not to mention blood pressure, cholesterol, heart rate, blood oxygen level, and more.

Obamacare takes the population medicine that Dr. Topol bemoans to the extreme. The law gives the federal Department of Health and Human Services authority to determine what, exactly, health insurance will cover. Secretary Sebelius famously exercised this prerogative in 2012 to mandate that all insurance policies include free contraception, but the authority reaches far beyond dispensing birth control. HHS bureaucrats will determine which treatments people get (i.e., what insurance will cover). As the *American Spectator* reports, the secretary of HHS will "determine what type of insurance coverage every American is required to have. She can influence what hospitals can participate in certain plans, can set up health insurance exchanges within states against their will, and even regulate McDonald's Happy Meals."[11]

Even more grotesquely, the law establishes the Independent Payment Advisory Board (IPAB) to make large cuts to Medicare without Congress's bearing any political pain. The board will determine what drugs, procedures, and therapies are covered and how much healthcare providers (doctors and hospitals) are paid for them. This board of unelected experts (who, once confirmed, essentially cannot be fired) will have an enormous effect on the medical treatment of millions of Americans on Medicare. And because Medicare is such a large force in the healthcare system, the IPAB will affect how the rest of us are treated as well. One central board of experts making decisions for an entire country on Medicare coverage? This is the epitome of population medicine.

Will Medicare cover the nanosensors that Dr. Topol predicts will warn of heart attacks before they happen? Will it pay for the blood glucose–level monitors for diabetics? Will it cover drugs that are customized just for you? The IPAB experts will decide. All we know for sure is that the board's ultimate aim is to cut costs, not necessarily to improve care.

In the end, bureaucratic healthcare and population medicine fail for the same reason: the so-called "experts" can never know enough about the needs of individual patients to provide them the best care, so they lump people into large classes and treat them all the same. That is to say, they often give everyone mediocre treatment.

There's another reason Obamacare threatens the breakout in medicine: its cost, both for people purchasing health insurance and for taxpayers. The House Energy and Commerce Committee recently reported that "consumers purchasing health insurance on the individual market may face premium increases of nearly 100 percent on average, with potential highs eclipsing 400 percent."[12] The Congressional Budget Office now estimates the law will cost taxpayers $1.8 trillion by 2023, double the original projection.[13] And that was before the administration announced it would offer subsidies to purchase insurance on the exchanges on what Reuters described as "the honor system" until at least 2015, because the government was incapable of implementing a verification system even three years after the law was passed.[14]

That cost will come at the expense of some real innovations like the ones Dr. Topol described. To help cover the legislation's enormous cost, Obamacare includes a tax on medical devices, a tax that the FDA, to no one's surprise, is now interpreting broadly. The agency plans to treat health-related mobile apps, of which there are already more than a million, as medical devices.[15] Smartphone-linked sensors will be regulated and taxed by the federal government. Free nutrition apps, WebMD, and even the Nike FuelBand could be as well if they make medical recommendations.

My friend Dr. Tim Rowe directs the Vertebrate Paleontology Laboratory at the University of Texas at Austin. When I visited recently, he showed me how his team, which used to spend hours extracting fossils from rock and cleaning them, can now simply CAT scan an embedded fossil in its crate, make an exact digital model, and reproduce it with a 3-D printer. The scientists end up with a perfect replica without ever touching the original specimen. (You can watch our conversation about this project at www.BreakoutUniversity.com.)

Tim mentioned in passing that doctors will soon use the same technology on persons who need hip replacements. They'll CAT scan a patient's hip, use the imaging to develop a digital model of the bone, and print out a replacement that perfectly matches the patient's hip joint. The medical applications of 3-D printing had not previously occurred to me, and they illustrate two useful concepts. First, Tim has been using this technology in his fossil lab for more than ten years—in fact, he launched an online database at Digimorph.com in 2002 to distribute his fossil models all over the world for free—yet the technology still has not made its way into most hospitals. Second, getting this clinically proved technology into hospitals will take even longer because of Obamacare. The technology will be regulated by the FDA, and the same machines will be taxed as medical devices.

★ ★ ★ ★ ★

Obamacare takes exactly the wrong approach to healthcare reform. Instead of encouraging medical breakthroughs, it hinders them and slows their adoption. The story of ten-year-old Sarah Murnaghan vividly illustrates the point. Her suffering is a preview of the cold and impersonal bureaucratic health system that Obamacare is only beginning to usher in.

Sarah was born with cystic fibrosis. By the time she was ten, she was on oxygen virtually around the clock. She desperately needed a lung transplant—preferably a pediatric lung, one suited to her size. Pediatric lungs are only rarely available for transplant, however, and by the spring of 2013, Sarah's doctors worried that she would not get a new lung in time to save her life. Her doctors decided, therefore, to approve her for an adult lung transplant. They were confident she would do well with a set of adult lungs pared down during the surgery. They added her to the adult transplant list.

Organ transplants are managed by the Organ Procurement and Transplantation Network (OPTN), established by Congress and overseen by HHS, which allocates available organs according to the urgency of a patient's need and his prospects for recovery. OPTN gave Sarah Murnaghan a high priority, an evaluation that should have put her near the top of the list and virtually assured her of receiving the transplant her doctors said she needed. There was only one problem: Sarah was fourteen months shy of her twelfth birthday, and twelve was the youngest age for eligibility for an adult transplant. When the OPTN system was created a decade earlier, in 2004, the number-crunchers felt there were not enough data to evaluate younger patients—so they were simply excluded from consideration.

The transplant bureaucracy therefore placed Sarah at the bottom of the list despite the urging of doctors at the Children's Hospital of Philadelphia (among the best pediatric hospitals in the world).

The OPTN policy was practically a death sentence for any child under the age of twelve in need of a lung transplant. No matter how well suited for an adult lung their doctors judged them to be individually, the abstract, population-level transplant bureaucracy would never allocate them an organ until every single patient on the list over the age of 12 had either refused it or received a transplant of their own. The under-twelve exclusion essentially said that because so few children needed new lungs, the children who required them couldn't have them.

It sounds crazy that official policy would deny medically sound organ transplants to ten-year-old children while giving them to older, less severely ill adults. For Sarah Murnaghan and children like her, however, it was deathly serious. According to a letter prepared by a lawyer for Sarah's family, 62 percent of the children on the waiting list in 2011 died before they got a transplant, compared with just 26 percent of the adults. If OPTN simply allowed children to be included in the adult rankings when their doctors judged an adult lung transplant medically appropriate, this disparity would have been far less dramatic.

The policy that was senseless and cruel in individual cases was no doubt the product of good intentions. The officials at OPTN, armed with statistics about adult transplants, crafted an algorithm to calculate, at the population level, the optimum allocation of scarce resources. Since the statistical evidence for children didn't meet their standards of significance, they excluded the group from the scoring algorithm. A policy that made perfect sense to a central bureaucrat whose job was to solve an optimization function left a ten-year-old girl with a dire need of a new pair of lungs and a prognosis for a successful surgery permanently trapped at the back of the line.

In the spring of 2013, Sarah's lawyer wrote to the HHS secretary, Kathleen Sebelius, asking her to suspend the under-twelve rule. Secretary Sebelius declined to exercise that authority, saying she wouldn't waive the pointless rule until it could be reviewed by the board of experts at OPTN. But Sarah didn't have that long to wait. Finally, in an unusual move, a federal judge intervened and ordered the rule set aside. Sarah got her lung transplant, and as her doctors had said all along, the surgery was successful.[16]

Systems that predictably produce policies like the under-twelve rule and stories like Sarah Murnaghan's are worse than misguided; they are immoral. They're flat-out wrong—but increasingly common.

Although Sarah's ordeal was not a product of Obamacare, it is a preview of the healthcare we can expect if the prison guards have their way. Impersonal and arbitrary bureaucratic rules—far removed from the judgment of individual patients and their doctors—will inevitably produce senseless suffering. The Obamacare law gave federal bureaucrats the authority to write thousands of rules that will determine everything from what treatments insurers will cover and how much doctors are paid to how the government handles your personal health information. And while rationing is unavoidable today with organ transplants, it will spread to other fields of medicine under Obamacare, and federal bureaucrats will continue to make rules as pointless and harmful as the under-twelve rule that almost killed Sarah Murnaghan. Sometimes the pernicious results of this bureaucratic rule-making will not be unintended. Interested individuals, industries, and ideologues will figure out how to influence the rules for their own purposes, regardless of what is best for patients.

The prison guards who write these rules and would like to write more of them argue that we just need to tweak this or that policy or add such and such an agency. They believe problems like Sarah's can be avoided with more money and more bureaucratic power. But "experts" can never know enough to make medical decisions for thousands or millions of people. The more they try, the more inhuman the health system will become.

Still, doesn't some public authority have to manage the allocation of scarce medical resources? As Secretary Sebelius pointed out in her comments on Sarah Murnaghan, *somebody* has to decide who gets a lung (and lives) and who doesn't (and dies). Within the current system, that might be true. But it doesn't have to be that way. Pioneers are working to change the current system. A breakout is on the way that could eliminate problems like the one Sarah faced.

Regenerative Medicine

Dr. Anthony Atala is working toward a day when Americans will not die while waiting for an organ donation. He envisions a world, much sooner than you might think, when instead of adding your name to a list and waiting for someone else to die, doctors will simply grow you a new organ—from you.

Atala is one of the world's leading doctors in the exploding field of regenerative medicine, which focuses on healing or replacing patients' organs or tissues using their own cells rather than using drugs or relying on organ donations. He is the director of the Wake Forest Institute for Regenerative Medicine, where researchers are achieving remarkable, almost unbelievable, breakthroughs. They are literally growing organs in the laboratory that can become functional in the body.

"Regenerative medicine is exciting because it really has the opportunity not just to manage disease, like a drug would, such as for someone with high blood pressure or diabetes, but really to cure it," Atala told me. "It really is a totally different approach for medicine."

The first organs Atala and his team regenerated to be put in real patients were bladders. They started by extracting cells from a patient's failing organs. Since each person's cells are unique, they harvested the cells that "knew" how to be part of that particular organ. They put the cells in an incubator that matched the conditions of the human body. The cells began to reproduce, and soon Atala's team had dishes full of new bladder cells belonging to their patients.

A dish full of cells—even a dish full of bladder cells—doesn't magically transform itself into a bladder. The cells need a shape to form around. So Atala and his team used a special biomaterial, similar to cloth, to stitch together a mold, or a "scaffold," in the shape of a bladder. This scaffold would give the loose cells something to grab onto as they cohered into a new organ.

The scientists carefully sucked up some of a patient's new cells with a dropper and coated the bladder-shaped fabric. Then they returned the scaffold to an incubator that mimicked the conditions of the human body. A few weeks later, laboratory tests confirmed that the cells were attached to the scaffold.

The scaffold was then transferred to the patient's body, where it continued to develop. Within a few weeks, the biomaterial would dissolve and only the complete, healthy bladder would remain.

This was the first time in history anyone had received a completely new organ grown in a lab. Today, Atala told me, "you have bladders that have been implanted into patients and we now have patients walking around for twelve years with their engineered organs." But bladders were only the beginning for Atala and his team. They used similar processes to grow replacement urine tubes that were implanted in patients, and today they are working on organs and tissues for more than thirty different areas of the body, including muscles, arteries, blood vessels, heart valves, kidneys, and livers. In some cases they actually "exercise" the moving parts (like valves and muscles) as they grow the tissues to condition the cells to life in the body.

Seeking to advance beyond the slow process of eyedroppers and fabric scaffolds, the team next discovered how to modify a standard inkjet printer, replacing the colored inks with different types of cells. Soon they built more-sophisticated printers and were 3-D printing pieces of bone in the lab as well as muscle and cartilage.

What they're working on today is even more astonishing. In a TED talk, Atala described the next generation of bioprinters, "where we print right on the patient.... you actually want to have the patient on the bed with the wound, and you have a scanner, basically like a flatbed scanner.... that first scans the wound on the patient, and then it comes back with the print heads actually printing the layers that you require on the patients themselves."[17] Today, Atala's team is using this combination of scanning and printing experimentally, to treat surface wounds such as severe burns. It's not hard to imagine that one day similar tools could perform regenerative surgeries on damage inside our bodies.

3-D printers can already do more than just build bones and heal skin wounds. Atala demonstrates a machine that takes a 3-D scan of a patient's kidney inside his body and digitally slices it up into thin layers. This information is used by a 3-D printer to "print" a prototype kidney by laying down scaffold material along with the patient's own cells.

Although these prototypes haven't been tested in humans yet, Atala's lab already has successfully implanted a mini-kidney in a steer that produced a urine-like substance. "We are growing kidneys in a laboratory right now," he said. "Miniature kidneys have been implanted into animal models that show that urine can be produced. So these things are real. These kidney structures can be created, they can be implanted, and they can function."

The breakthroughs that Dr. Atala and others like him are pioneering could lead to the most important advances in medical care in generations. Many of the most difficult, most painful, most expensive, and most lethal health problems could simply cease to exist if we could routinely regenerate organs.

If your kidneys are failing today, you require dialysis, which takes many hours per day. Tomorrow, your doctor may instead print new ones. Today, nine out of ten patients waiting for a transplant are waiting for a kidney. That's more than ninety-three thousand people. In 2011, almost five thousand people died waiting.[18] Moreover, 355,000 Americans are on dialysis,[19] therapy that costs taxpayers $20 billion a year. The annual cost of dialysis is a quarter of a million dollars per year per patient, according to Dr. Atala. A lab-grown kidney would be a bargain in comparison and of course would provide an enormous improvement in quality of life.

Are you one of the nineteen million Americans who suffer from diabetes?[20] Instead of living with a chronic disease for decades, you could one day get a new pancreas grown by your doctor. Since diabetes costs $245 billion every year, these too might be a bargain.[21] Certainly, they'd be lifesaving for the 1,400 people now waiting for a transplant.[22]

Children like Sarah Murnaghan, too, could simply wait a few days for their doctors to grow new sets of lungs from their own cells instead of waiting for a transplant that might never come. The shortage of organs today is severe. More than 119,000 people are in line for some kind of transplant in the United States. Eighteen people die every day waiting for an organ.[23]

Many types of cancer, too, could be cured by replacing the affected organs with lab-grown alternatives: lungs, pancreases, breasts, colons, livers, and more might simply be swapped out for non-cancerous versions.

Organ regeneration could eliminate many of the leading causes of death, giving us longer and healthier lives. Like a car in need of maintenance and some replacement parts at 150,000 miles, many persons might keep on going for decades with a little retooling.

★ ★ ★ ★ ★

There are still many scientific hurdles to overcome before Dr. Atala's pioneering work results in the mass use of regenerated organs. Atala estimates that most of what he demonstrated is some years away from being ready for prime time. But in addition to the scientific challenges, the process of getting breakthroughs past the FDA is astonishingly slow. We are likely many years away from seeing approvals for most of these therapies under the current regulatory model.

Standards for clinical trials were designed for the pharmaceutical industry, and those standards pose an extremely steep challenge for regenerative medicine. The FDA has major divisions for medical devices and for "biologics" (covering, for instance, vaccines and transplants). But regenerative medicine falls into both of these categories, Dr. Atala told me, which means that "the regulatory path is one that has to fulfill the requirements from both branches, which is really very complex." (Of course, as we saw in the case of Abigail Burroughs, it's hard enough to fulfill the requirements of one branch.)

In the trials the FDA currently demands, a company attempting to bring this technology to market would have to pay these huge costs on its own—and not just the cost of the technology. In addition to the extraordinary costs of growing an organ, clinical trials of regenerative medicine would have to cover surgical costs, X-ray costs, hospital costs, and inpatient costs, which are not common with other types of trials.

Testing regenerative medical technologies is not only expensive, it is fundamentally different from testing a pill. For one thing, the results are basically black and white. A new kidney either works or it doesn't. This is more definitive than a drug test, in which researchers observe patients for months to look for marginal improvements over other products on the market.

In pharmaceutical trials, every pill is identical. That's what makes the kind of testing the FDA demands statistically meaningful. But with regenerative medicine, as another doctor reminded me, "you're using the patient's own cells, so every time you're creating a product, you're really creating a different product because it's unique to that patient."

Many in the field agree that the FDA's unreasonable hurdles are keeping medicine trapped in the past, but the same doctor told me his colleagues are afraid to speak up because the agency holds the keys to everything they do. After all, the FDA has to approve every drug and every medical device a company might want to market, and no one wants to make the prison guards angry. They could draw out the process, demand more expensive trials, set impossibly high evidentiary standards, or even leak rumors to tank a company's stock price.

"We bring in people from industry, and they're worried to make the FDA upset, so they're actually afraid of making any suggestions for change," the doctor commented. "Everyone is afraid that they're going to get pounced on because they're making suggestions for change. They have too much riding on the stuff that they're doing right now, which is under the current regulatory process."

To avoid the time and expense involved in testing regenerative medicine products under current FDA regulations, many companies are simply fleeing the United States. "Yes, it is accurate that there are other countries where you can go fairly quickly to the clinic and try these technologies," Dr. Atala said when asked about this. "What you're seeing is there are many technologies that are being done abroad that are not yet done in the U.S., and that a lot of these breakthrough technologies are being done clinically first abroad rather than in the U.S."

How quickly could regenerated organs be widely available in the United States? Dr. Atala predicts that some organs could be available "within five to ten years" under a pathway designed specifically for the new field of regenerative medicine. How long will it take under the current FDA? Data shows that the average regulatory path for a pill—just for a pill—from the time it reaches the first patient of a clinical trial to the time it actually gets distributed to the open market is 15.1 years.

It wasn't always this way, one doctor tells me. "The process used to be short. It did not take fifteen years before. There was a time when the process only took six years. And there was a time when the process went up to eight years. Then it went up to ten years. In the 1990s, it was twelve years. Now it's fifteen years."

Nearly a quarter of all products Americans consume are regulated by the FDA. Every single drug and medical device we rely on must pass the inspection of FDA bureaucrats, not to mention nearly 80 percent of our food supply. That's an extraordinary amount of power for one federal agency.

Americans deserve an FDA that protects consumer safety but also makes sure lifesaving breakthroughs get from our labs to our pharmacies and hospitals as efficiently as possible. That's far from what we have today. The agency is still clinging to an obsolete, decades-old regulatory model that will drown the strides in regenerative medicine that could enable us to treat kidney failure, diabetes, many cancers, and other diseases with a brief trip to the hospital. Without a major transformation, the FDA will keep American patients trapped in the medicine of the past, probably for decades.

Think that's an exaggeration? Consider how long it has taken another lumbering federal bureaucracy, the Federal Aviation Administration, to approve Kindles and iPods for use on commercial airplanes below ten thousand feet. Fifteen years after the introduction of the first MP3 players and nearly a decade after e-book readers became widely available, passengers are still prohibited from reading on Kindles or listening to music for much of their flights. In fact, the entire product life of the iPod has come and gone and e-books have surpassed print books as the dominant format for publishing without the FAA's noticing. There is no evidence to justify the ban on these items; millions of passengers routinely flout the rules with no effect on safety. The FAA simply hasn't updated its regulations on in-flight electronics since the 1960s. The agency, moreover, is still mired in the effort to convert the air traffic control system from 1950s-era radar to GPS, which has been commercially available for fifteen years and which virtually all passengers carry onboard the plane in their pockets.

The FDA is as slow and resistant to change as the FAA, but delays and rejections of new medical technologies are a matter of life and death. Congress gave the FDA authority to judge the *efficacy* of drugs, in addition to its original mission of assessing *safety*, in 1962. Evidence suggests this new layer of bureaucracy caused a dramatic decline in the number of new drugs introduced. The *Wall Street Journal*'s Daniel Henninger reports, "A 1974 study by University of Chicago economist Sam Peltzman concluded that since 1962 the new rules had reduced the rate of introduction of effective new drugs significantly—from an average of forty-three annually in the decade before the amendments to just sixteen annually in the ten years afterward."[24] That's a 63-percent decline in the number of new drugs that reached pharmacies in the decade following the agency's expansion. Delays continued into the next decade. Henninger cites a study from the Tufts University Center for the Study of Drug Development that concluded that nearly three-quarters of the forty-six drugs approved by the FDA in 1985 and 1986 were available earlier in foreign markets, by an average of five and a half years.[25]

The FDA's onerous regulations and strict clinical trial standards come at a price. Avik Roy of the Manhattan Institute reports that development costs rose from $100 million per drug in 1975 (2012 dollars) to $1.3 billion in 2005—a thirteenfold increase.[26] The bulk of this increase comes from the exploding cost of the phase III clinical trials, which the FDA scrutinizes heavily. In the six years from 1999 to 2005, Roy reports, "the average length of a clinical trial increased by 70 percent; the average number of routine procedures per trial increased by 65 percent; and the average clinical trial staff work burden increased by 67 percent."[27]

No wonder we get the one-pill-fits-all medicine Eric Topol laments! How could any company personalize its drug for *you* if it had to spend hundreds of millions of dollars on a two-thousand-patient clinical trial for the FDA?

Needless to say, startup costs in excess of a billion dollars are a colossal barrier to innovation. Only the very largest pharmaceutical companies can afford to bear them, and they can spend such sums on only a handful of major drugs. It is almost unthinkable that an independent business with a technology like Dr. Atala's could survive such a process.

Even if the exorbitantly expensive and redundant trials the FDA demands inhibit innovation, aren't they there to keep us safe? Safety may be the goal, but deterring innovation is deadly. Ronald Trowbridge and Steven Walker of the Abigail Alliance for Better Access to Developmental Drugs (named for Abigail Burroughs, whom we met in the introduction to this book) recount one outrageous example:

> Beginning in June 2004, we started pushing the FDA to make Nexavar and Sutent, both highly promising drugs for kidney cancer, available. The agency eventually approved Nexavar in December 2005 and Sutent in January 2006. But that was only after evidence of efficacy so compelling emerged for Nexavar that the trial demanded by the FDA—in which dying kidney cancer patients seeking the drug were being given no other choice (except certain death from their cancer) but to agree to a 50/50 chance of being blindly randomized to a sugar pill—was stopped by Bayer for ethical reasons and the placebo patients allowed to get the drug. The sponsor seeking approval for Sutent was given a similar option by FDA if it wanted its drug approved. About 20,000 kidney cancer patients died waiting for both drugs.[28]

Trowbridge and Walker cite numerous other delays that denied tens of thousands, sometimes hundreds of thousands, of dying patients the breakthrough drugs that might have saved their lives. Yes, the FDA might grant a handful of these patients the "compassionate use" waiver that

Abigail sought, but drug companies themselves are understandably reluctant to go along lest they endanger their billion-dollar investments in tightly controlled clinical trials.

If this is how the FDA handles new pharmaceuticals, what will it do to the transformational new technologies of regenerative medicine? The prison guards could set us back decades. Since the FDA is obviously unprepared to adapt to such breakthroughs and speed them to market as efficiently as possible, the American people should demand the right to opt out of this lethal system. Champions of the coming breakthroughs should be able to bypass the FDA with informed consent. Patients whose doctors advise them to accept a lab-grown bladder or a new heart valve should be able to waive the prison guard seal of approval with a written acknowledgment of the risks. The chances are that many of us will want the breakthrough anyway.

It is true that this freedom might pose a challenge to the old model of clinical trials in which half of the patients get placebos and half get the actual drug. After all, it might be harder to get patients to agree to those odds if they can sign a waiver and be sure that they are getting the breakthrough treatment. But with continuous monitoring, as Dr. Topol demonstrated, and decades' worth of clinical research data under our belts, it is time to look for more humane models of testing, anyway.

Winning this freedom will not be easy. The Blue Diamond nut company got a taste of the prison guards' insane jealousy when it received an enforcement letter from the FDA. Blue Diamond's labels had touted the natural health benefits of some of its products. For instance, one stated, "The omega-3 in walnuts can help you get the proper balance of fatty acids your body needs for promoting and maintaining heart health." The FDA warned Blue Diamond, "We have determined that your walnut products are promoted for conditions that cause them to be drugs because these products are intended for use in the prevention, mitigation, and treatment of disease." Blue Diamond walnuts were not only drugs, but they were *new* drugs that had yet to be approved by the FDA and therefore could not legally be sold. Since the "walnut products are offered for conditions that are not amenable to self-diagnosis and treatment by individuals who are not medical practitioners," the FDA letter continued,

"... adequate directions for use cannot be written so that a layperson can use these drugs [the nuts] safely for their intended purposes."[29] In other words, even if the FDA "approved" the company's walnuts as a drug, they would require a prescription from a doctor. The company surrendered and changed the labels.

Let anyone who thinks the prison guards will relinquish their power without a fight be instructed.

Americans could soon enjoy longer and healthier lives with personalized medicine and regenerative technology. But the prison guards could delay those benefits by decades. Tens of millions of Americans don't have that long to wait.

CHAPTER FOUR

BREAKOUT IN AMERICAN ENERGY

After dropping out of high school at the age of sixteen, Andy Turco took work wherever he could find it. Staining decks in the summer and shoveling driveways in the winter, Andy was willing to work hard, but steady employment was difficult to find. He often went months between jobs, and even when working, he earned so little that he was often on the brink of destitution. Surviving was a struggle. Actually improving his life seemed impossible.

That is until he talked to a friend who had recently gotten a great job in a booming industry. For Andy Turco, a personal breakout to a better life was not impossible after all.

It was a pipe dream—an oil and gas pipe dream.

Today Andy makes nearly six figures working ninety hours a week on a drilling rig in the remote town of Williston, North Dakota.[1] The story of how Andy and millions like him all across America have achieved a better life thanks to amazing innovations in oil and gas production is a perfect illustration of the great breakout that is possible in America. It begins with a pioneer who ignored conventional wisdom to develop a technological breakthrough nobody else thought possible.

More than thirty years before Williston became a boomtown, President Jimmy Carter held a televised conversation with America, one he warned would be "unpleasant." Our country, he said, was running out of oil and natural gas. The solution, according to Carter and the conventional wisdom of the time, was for the United States to reduce radically the amount of energy it used.

Laying out a plan to conserve our "rapidly shrinking resources," the president used the word "sacrifice" ten times. And the energy policy built upon a foundation of pain and sacrifice succeeded in creating just that—gas shortages, high energy prices, and economic stagnation.

All of this suffering was totally unnecessary, but the president and his advisors were prisoners of the past—trapped by a prison guard as dangerous and difficult to defeat as teams of lobbyists or intractable bureaucrats: a false idea accepted as conventional wisdom.

President Carter and most of the energy experts at the time adhered to the "peak oil" idea. The prevailing theory of the time, it stemmed from a 1956 study by M. King Hubbert, who concluded that patterns of production for any finite resource follow a predictable pattern. Hubbert's model forecast that the United States would reach its peak oil production in 1970 and decline gradually from there.

Many "experts" considered the peak oil theory validated after production indeed reached a plateau in 1970 and began falling steadily after 1985. The problem with the theory of peak oil, however, is the same one that plagues all projections of global disaster and is common to all prison-guard ideas: it fails to account for the breakthroughs by which pioneers solve our toughest problems and create new opportunities.

The peak oil projections did not properly account for the rate of new oil and gas discoveries and improvements in technology for extracting them. The truth was that we still had enormous resources in America—we simply lacked the means to obtain them at a reasonable cost.

In a sense, peak oil and the other theories that put limits on opportunity and growth are pessimistic theories of "peak America." They can be dangerously self-defeating.

One of the people who understood this was George Mitchell. While the stated policy of the country was to move away from natural gas as a source of energy, this son of Greek immigrants staked his company on a countervailing view—that vast amounts of natural gas were trapped deep underground, just waiting to be safely and economically tapped.

Shale is an extremely dense rock, often between one and two miles below the earth's surface. At that depth, the rock traps individual molecules of natural gas and oil between its thin layers. Geologists had known this for decades, although they did not appreciate the scale of the resources—and more importantly, nobody had figured out how to extract the oil and gas at a reasonable cost.

Starting in the early 1980s, Mitchell drove his company to develop new technologies for releasing the gas trapped in rock thousands of feet below ground near Fort Worth, Texas, in a geological mass known as the Barnett Formation. Progress was slow and costly. Mitchell's engineers knew that creating fractures in the shale was the key to extracting the gas. Typically, this involved injecting a fluid into the well at sufficient pressure to push its way into small fissures in the rock below ground, freeing the molecules of oil or gas trapped inside. In fact, developers had been using "fracking" techniques to extract more resources from wells for decades. But nobody had done it on the scale Mitchell had in mind or in such dense rock.

Mitchell's company spent a decade working to extract gas from the Barnett Shale, but its wells cost much more than traditional ones, and the company ran short on money.[2] Then three big breakthroughs changed Mitchell's fortunes and in the process changed the world.

The first came when an employee at the company suggested adopting a much less expensive fracking technique from a rival company. Instead of pumping a costly gel-based fluid into the rock to create fractures, they used mostly water, combining it with sand to prop open the tiny cracks.

The mixture was only about one-fifth the cost of the gel, and it ended up yielding more gas.[3]

The second breakthrough came soon afterward. A geologist named Kent Bowker suspected that there was much more gas trapped in the Barnett Shale than was thought, but he needed funding for the study. When he brought his proposal to Mitchell Energy, the company president, Bill Stevens, turned him down. Fortunately, Mitchell himself heard about Bowker's proposal and overrode the decision. The astounding conclusion of Bowker's study was that there was four times as much gas in the Barnett Shale as previously estimated.[4]

Finally, Mitchell convinced the Department of Energy to cover part of the cost of an early horizontal well in the area, an impressive engineering feat he hoped would help extract the enormous supplies more efficiently.[5] The technique involves drilling a standard well straight down into the dense, high-pressure shale rock, but then rotating ninety degrees and drilling outward almost two miles. Fracking a well that drives horizontally through the thin layers of shale increases enormously the amount of gas that can be extracted from a single point on the surface. Indeed, horizontal drilling can often obtain more gas with one well than a driller might obtain with dozens of traditional wells. This finally made it economical to free the huge resources trapped in the underground shale. The breakthrough technology was better for the environment, too, since fewer surface wells were needed.

After nearly two decades of work, George Mitchell's vision of extracting natural gas from the Barnett Shale was a reality. It didn't take long for other energy pioneers to figure out that Mitchell was doing something right. Across the country—in Texas, Louisiana, Colorado, Pennsylvania, Ohio, and New York—they began to apply Mitchell's innovations to obtain energy in places no one had ever dreamed of trying. Most famously, Harold Hamm and his Continental Resources applied Mitchell's fracking techniques to get oil from the Bakken Shale in North Dakota, beginning the boom that provided Andy Turco and tens of thousands of others with lucrative new jobs.

More importantly, the breakthrough shattered the peak oil myth, and with it, the excuse for a high cost of energy and the rationale for the painful policies left over from the Carter era. Thanks to the combination of technological breakthroughs and new shale discoveries, by the end of the first decade of the twenty-first century, U.S. production of oil and natural gas was rising for the first time in forty years, and the nation was on the verge of a breakout in American energy.

No Limit

In the years since George Mitchell and Harold Hamm discovered an affordable way to free molecules of oil and gas trapped in rock thousands of feet below ground, we have learned that America's energy potential is practically without limit. Even while many Americans—especially those in Washington and in the media—continue to claim we are running out of oil, estimates of the supplies we could obtain right here in the United States keep going up.

According to the Institute for Energy Research's *North American Energy Inventory*, the United States has 1.4 trillion barrels of recoverable oil and 2.74 quadrillion cubic feet of natural gas—an almost incomprehensibly large supply. To put it in perspective, that's enough to:

- Fuel every passenger car in the United States for 430 years
- Provide the United States with electricity for 575 years at current natural gas consumption levels
- Fuel homes heated by natural gas in the United States for 857 years[6]

Half a millennium's worth of American oil and gas should be enough to put even the most alarmist mind at ease, but even these figures are likely too conservative. Our estimates of recoverable oil and gas reserves increase every few months, and the continual breakthroughs in drilling technology give us access to resources that were once technically or economically unreachable.

Estimates of oil and gas reserves consistently increase once a site is explored. Indeed, officials from the Energy Information Agency have admitted that the agency's estimates usually prove low.[7]

In 1977, for example, when production first began in Alaska's Prudhoe Bay, the official reserve estimates were nine billion barrels of oil. Yet by 2012, the area had produced fifteen billion barrels—66 percent more than expected in the 1970s, and production is still climbing. In 1984 the Minerals Management Service estimated the offshore land in the Gulf of Mexico held six billion barrels of oil and sixty trillion cubic feet of natural gas. Since then, we have produced more than twice those amounts, and production is still climbing.

The most dramatic increases have occurred where technological breakthroughs have magnified our capabilities. For instance, no one used to think that the Marcellus Shale in Pennsylvania, New York, Ohio, and West Virginia held much gas worth extracting. The U.S. Geological Survey estimated as recently as 2002 that the formation had about two trillion cubic feet of recoverable natural gas. Just ten years later, however, the USGS revised its estimate to eighty-two trillion cubic feet—an extraordinary 4,100-percent increase.[8] Even this estimate is on the low end. Professors at Penn State and the City University of New York, Fredonia, now estimate that Marcellus could hold an astounding five hundred trillion cubic feet of recoverable natural gas.[9]

Perhaps the best example of how our understanding of America's energy resources is constantly changing is how far off the initial estimates of the Bakken Shale in North Dakota proved to be. In 1995, the USGS estimated that the formation held just 150 million barrels of technically recoverable oil. As drilling technology improved and pioneers like Harold Hamm explored further, however, the USGS revised its estimate upward to four *billion* barrels in 2008—a twenty-five-fold increase. Even that figure pales in comparison to the latest estimate from Hamm's Continental Resources, the largest drilling operation in North Dakota. Continental now forecasts a whopping twenty-four billion barrels of recoverable oil in the Bakken Formation—160 times (16,000 percent of) the initial estimate.[10]

Remarkably, the latest estimates of the Bakken may be just scratching the surface. As Stephen Moore of the *Wall Street Journal* reported recently, the amount of recoverable oil in the formation could exceed five hundred billion barrels as drilling technology continues to improve.[11]

New oil and gas discoveries and technological breakthroughs are occurring at such an impressive rate that by the time you read this book, many of these estimates will likely be out of date. As Nansen Saleri of Quantum Reservoir Impact put it in a recent interview, in the coming decades we will look back at today as the "initial dinosaurish phase of shale and unconventional resource development."[12]

Thanks to oil production from shale, we are truly in a new world of American energy—a world that almost no one expected, but one that offers exciting opportunities for all Americans.

A Better Life for All Americans

We have already begun to see big changes from the explosion of oil and gas potential.

Take, for example, North Dakota, sitting atop the Bakken Formation. The story of Andy Turco, whom you met at the beginning of this chapter, is far from unique. Five years into the Obama presidency, when the national unemployment rate was stuck at around 7.5 percent, North Dakota's jobless rate was just 3.0 percent, thanks largely to the energy boom.[13] Tens of thousands of people with only a high school education have moved to North Dakota for jobs in the industry with starting salaries in excess of a hundred thousand dollars a year.[14] NBC News recently reported from Williston, North Dakota, that "there is such a large influx of people that thousands are staying in 'man camps'—shipping containers converted into housing units for the workers new to town. When more teachers were hired to deal with the rising number of students, an apartment building had to be built to house the new teachers."[15]

While other states were struggling with fiscal crises, raising taxes, and cutting services, North Dakota was actually cutting taxes because of the massive influx of revenue from its booming energy economy.

Similarly huge discoveries of natural gas in the Marcellus Shale spurred a wave of growth and opportunity in Pennsylvania, Ohio, and West Virginia, a region that had been written off as the "Rust Belt." The onetime backbone of American manufacturing had been decimated in recent decades as factories moved overseas, but finally, after many years of decline, new jobs and investments are rushing into the area.

Pennsylvania, one study showed, got sixty-seven thousand new jobs related to natural gas development in 2010 alone, including 23,000 jobs in construction, 13,600 in mining, and 1,900 in hotel and food services. The study found that the "total employment impact" of natural gas development that year was 140,000 jobs.[16] Pennsylvania's Department of Labor estimates that the average salary for jobs in shale-related industries is $73,000 per year—$23,000 more than the average in all industries.[17] Meanwhile, one gas well in Susquehanna County has provided more than a million dollars in royalties to the Elk Lake School District,[18] and a Dutch petrochemical company recently announced plans to build a two-billion-dollar plant in Beaver County.[19]

The American energy breakout is not only benefiting the places fortunate enough to produce oil and gas. It is improving the lives of Americans throughout the country. A recent study by IHS Global Insight showed that hydraulic fracturing supported 1.7 million jobs in 2012, including many workers outside the energy industry. This number is projected to increase to 3.5 million by 2035.[20]

Moreover, it has caused an astonishing drop in the cost of energy nationwide. In 2008, the average price of natural gas was $8.86 per million BTUs. In 2011, it was four dollars. In 2012, it was just $2.75. In April 2012, the spot price for natural gas hit a thirteen-year low at just $1.95, an 80-percent drop from the same month in 2008. By the summer of 2013, the price had rebounded a little to $3.62, but the decline is still staggering.[21] It is leading to a lower cost of energy for all Americans and powering a renaissance of manufacturing here in the United States.

The American energy breakout has also made us safer. A record 60 percent of our oil was imported in 2005. Six years later, that figure had dropped to 45 percent thanks largely to new domestic production.[22] The Energy Information Agency now predicts that the United States will replace Saudi Arabia as the world's leading petroleum producer by 2017 and will become a net oil exporter by 2030.[23]

The geopolitical implications of this change cannot be overstated. For decades the members of the Organization of Petroleum Exporting Countries, many of which are dictatorships, have been able to blackmail Western countries with their power to manipulate oil prices. As oil production in the United States (and Canada) increases, the influence of OPEC falls.

This shift hasn't been lost on the Saudis. In the summer of 2013, Prince Alwaleed bin Talal warned his country's oil minister that demand for OPEC oil was "in continuous decline" and that the kingdom's economic dependence on oil was a "truth that has really become a source of worry for many."[24]

"Our country is facing a threat with the continuation of its near complete reliance on oil, especially as 92 percent of the budget for this year depends on oil," he continued.[25]

Much of the money the West pays Saudi Arabia for its oil ends up sponsoring some horrible activities. The kingdom is the top funder of Wahhabism, a virulently anti-Western, anti-Christian, and anti-Semitic version of Islam, and Saudi-based "charities" remain one of the top sources of funding for Al Qaeda and the Taliban.[26] Thanks to the shale oil and gas breakthrough, fewer and fewer of our dollars are going to support our enemies.

Peak Oil: A Self-Fulfilling Prophecy

The breakthroughs in American energy might already have produced some impressive results, but the broad breakout we could see based on this huge potential is still far from inevitable.

A straight line that runs through the Marcellus Shale illustrates why. On one side of this line, energy companies are paying local residents thousands of dollars per acre for the rights to drill on their properties. Many families have enjoyed huge paydays as the new technology has come online and developers have competed for mineral leases. On the other side of this line, the shale is much the same—full of natural gas just waiting to be released. Above ground, however, the region's major energy company is packing up and leaving town, abandoning leases it made years ago for as little as three dollars an acre.[27]

What is this line? It's the New York–Pennsylvania border. South of the border, Pennsylvania counties are prospering in the fracking revolution, seeing jumps in per capita income of nearly 20 percent since 2007. Their neighbors to the north are shut out of the party by a state government that continues to ban the technology despite ample evidence of its safety.

The artificial restrictions on our energy potential are perhaps most vivid here, but similar prison-guard policies are restricting energy production throughout the country. In fact, a large portion of the resources I have mentioned—the supplies that could fuel every passenger car in the United States for 430 years—are on federal lands owned by the American people. But rather than leasing them for development and putting downward pressure on the price of gasoline, prison guards in the government have kept them under lock and key.

The question of whether America can break out in energy is almost entirely dependent on government policy and not on the amount of oil and gas within our borders.

While oil and natural gas production potential was exploding in the first decade of the twenty-first century, most of our leaders in Washington were oblivious. A full six years after George Mitchell made his big breakthrough, Congress passed legislation to speed the construction of liquefied natural gas terminals for importing—not exporting—the hundred billion dollars per year of natural gas it thought we would need for electricity.

I have to admit that while I was bullish on America's capacity to produce more oil, it was a while before I appreciated the magnitude of the shale oil and gas revolution. In fact, in my 2008 book, *Drill Here, Drill Now, Pay Less*, I advocated steps to diversify the ways the United States produced electricity, expecting a shortage of natural gas.

This ignorance of the energy revolution occurring beneath our feet was not limited to Washington. Most energy "experts" were unaware as well. In 2008 the International Energy Agency predicted that the planet's oil production would begin to decline permanently by 2030. The United Kingdom Energy Research Centre predicted the decline would begin even sooner, by 2020.[28] Then in 2010 the IEA announced that we had reached peak oil in 2006 and that crude oil production would flatline until 2030 and then decline. The Institute for Policy Studies summed up this conventional wisdom in an article that declared, "It is the beginning of the end of the Petroleum Age."[29]

Even those who rejected the notion that we were running out of oil nevertheless thought we were reaching a maximum capacity of oil production. In late 2007 the *Wall Street Journal* reported, "A growing number of oil industry chieftains are endorsing an idea long deemed fringe: The world is approaching a practical limit to the number of barrels of crude oil that can be pumped every day."[30]

The survival of the peak oil myth long after the fracking and horizontal drilling breakthroughs is a perfect example of the power of the prison guards of the past to embalm bad ideas as conventional wisdom. Worse, it proves that conventional wisdom can become a self-fulfilling prophecy.

Even before the fracking and horizontal drilling breakthroughs, there was mounting evidence that the peak oil model was wrong. While Hubbert had correctly predicted that U.S. production would level off in 1970, his model badly miscalculated global oil production. Hubbert estimated global peak oil to occur in the year 2000 at just under forty million barrels of crude oil per day.[31] By 2000, however, the world was already producing 68.5 million barrels per day, and production has continued to rise since then. In 2011, the world produced seventy-four million barrels of oil per day.

So why were the peak oil predictions so far off when it came to global oil production but accurate about U.S. production? Why did U.S. oil production peak in 1970 and then decrease so substantially after 1985? The evidence suggests that federal energy policy based in part on the peak oil model, not dwindling supply, led to decreased domestic production of oil and gas.

While crude oil production in the United States fell steadily from 1985 until 2008 (about when the Bakken in North Dakota came on line), global production rose—growing to over 73.5 million barrels from 54 million barrels. Furthermore, estimates of the oil we *could* recover in the United States rose steadily—increasing to 166 billion barrels from 143 billion in

1996—even as actual production declined.[32] That discrepancy suggests that the forty-year stagnation and then decline in U.S. oil production had more to do with U.S. energy policy than with the amount of recoverable oil in the United States. The government had put so many restrictions on energy development that production collapsed as existing wells were depleted and too few new ones took their place.

In my 2012 book, *$2.50 a Gallon: Why Obama Is Wrong and Cheap Gas Is Possible Now*, I lay out in extensive detail the many barriers to U.S. oil and gas development. Here are a few highlights (or lowlights, as the case may be):

- For more than thirty years, the government has banned production in an energy-rich area of the Alaskan National Wildlife Refuge (ANWR) that the EIA predicts could boost our production of oil and gas by 20 percent. The debate about whether to permit drilling for these enormous resources centers on an area that amounts to less than 8 percent of the ANWR and just 0.3 percent of the total land area of Alaska (by far the most sparsely populated state). The ban endangers the Trans-Alaska Pipeline, a vital artery for American energy, which will need to be shut down completely if the amount of oil flowing through it decreases any further.
- Offshore, the government has imposed de facto moratoriums on energy development. The U.S. Outer Continental Shelf comprises hundreds of millions of acres underneath the ocean. The American people own this land, but the bureaucrats and elected officials in Washington have refused to allow the vast majority of it to be explored. The Minerals Management Service (now the Bureau of Ocean Energy Management) estimated in 2006 that the OCS contains lots of energy resources—86 billion barrels of oil and 420 trillion cubic feet of natural gas—but the true numbers are likely to be dramatically larger.[33] Today, however, the government blocks all but 2.4 percent

of the area from oil and gas production, while the unleased acreage is ten times the size of Texas.[34]

- Onshore, the federal government has drastically cut back on new mineral leases on the American people's land. In fact, the rate of newly leased federal lands contracted nearly 90 percent to fewer than two million acres per year in 2010, down from more than twelve million acres in 1988. The federal government is far from running out of land to lease for energy production, however. Out of all the undeveloped oil remaining on federal lands, 62 percent is inaccessible, according to a 2011 Interior Department study.[35] Fully 91 percent of undiscovered resources on federal lands onshore are either inaccessible or restricted.[36] So the drop in leases is due not to a shortage of land to lease but to government policy reducing the number of permits allotted and a bureaucratic process so complicated that energy developers have opted to pursue development on private land. It takes more than three hundred days to get a permit to develop on federal land, compared with just a few weeks in the states where energy production is booming.

  This restriction amounts to completely irresponsible management of the public's land, robbing the American people of jobs and money that could have been kept at home instead of sent overseas. The government is ignoring a huge source of non-tax revenue, too, which could be used to reduce our federal budget deficit. A study by Noble Royalties determined that bringing federal leasing back up to 1988 levels would generate an additional $421.3 billion over the next thirty years. The failure to responsibly develop the natural resources on federal lands is a hidden tax on the American people.[37]

Apart from outright bans on development and the failure to issue new leases, a regulatory assault spanning three decades has increased the burden on energy producers, making it difficult to develop American resources on either public or private land. In the next chapter, we will see how the aggressive regulatory state, pushed by a movement of "green" prison guards, is fighting with all the means at its disposal to limit the energy future Americans could enjoy. One small example: Excessive regulation is the reason the United States has not built a major new refinery since 1976. Yet without increasing our capacity to refine crude oil into gasoline, all the new supplies that pioneers like Harold Hamm have discovered won't make it to the gas pump, won't drive down prices, and won't create jobs.

My friend Scott Noble, the founder and president of Noble Royalties in Dallas, came up with a useful way to describe the effects of a hostile regulatory regime on energy development (or any other sector of the economy). He calls it "regulate-restrict-destroy."

Ask yourself if most of the regulations we have seen in this chapter (or indeed, in this book) ensure that an economic activity is conducted safely, or if they are designed to restrict or even eliminate the activity itself.

It's clear that the effect of U.S. energy policy in the past three decades has been to restrict and destroy energy development rather than enable it to be done safely. The fracking revolution occurred on private lands, avoiding some federal restrictions, but even private development is coming under assault. As we will see in the next chapter, this "restrict and destroy" energy policy is designed primarily to please radical environmentalists, but it was given a veneer of nonideological practicality by the peak oil myth. If we were about to run out of oil and gas anyway, the argument went, then we needed to hasten the transition to new sources of energy. Continued oil and gas development would make that transition harder.

So federal energy policy, which was based in part on peak oil theory, caused the peak oil model to come true—at least for a time. The resulting diminished production, in turn, reinforces the peak oil model and makes it even more difficult for us to break out.

Today, that mindset and the policies it justified are obsolete. The world has changed, and we must break out of the prison of that costly, false idea. But as we are about to see, the prison guards seeking to destroy the breakout in American energy aren't giving up without a fight.

CHAPTER FIVE

THE GREEN PRISON GUARDS

Restrictions for Reptiles

The University of Texas and the Texas A&M University systems are the beneficiaries of an endowment of $30 billion.[1] Among university endowments only Harvard's is larger. This vast wealth keeps tuition at the systems' flagship universities, UT Austin and Texas A&M, at about $10,000 a year for state residents.

The university students, professors, and administrators of Texas owe their good fortune to the transfer of a million supposedly worthless acres to the universities in 1883. That land was in a region of west Texas known as the Permian Basin, which a few decades later became the largest onshore oil-producing region in the United States.

The Permian Basin is also home to a tiny brown reptile known as the dunes sagebrush lizard. No more than three inches long, the lizard lived unnoticed in the desert sands until recent decades. Not long after the unremarkable creature was classified as a species of its own in the 1990s, it attracted the keen interest of a small group of academic researchers.

In 2006 and 2007, graduate students and research assistants from Texas A&M's Department of Wildlife and Fisheries Sciences began traipsing through the dunes, looking for the sagebrush lizards. Their aim, apparently, was to document where the species lived. But actually spotting them was difficult, since they are only about the size of a human thumb and are camouflaged to blend in with their surroundings. The sand dune lizards are flighty, too, known to run away when danger approaches. Even if the lizards were there, it was nearly impossible to find them.[2]

The graduate students visited twenty-seven sites among the west Texas dunes, and after looking around a bit, spotted the lizards at only three of them. The researchers' method for determining the presence of lizards—stomping around the dunes and asking, "See any lizards?"—hardly seems a model of scientific rigor, yet it led to a conclusion that the population had declined.

About six months after the group published the results of its "study," environmentalists in New Mexico petitioned the Interior Department to declare the dunes sagebrush lizard an endangered species.[3] By 2010, the Fish and Wildlife Service, relying on the graduate students' report, proposed adding the reptile to the list of endangered species.[4]

Species' populations fluctuate for a variety of reasons, many of them natural and often difficult to identify. But the Fish and Wildlife Service proposal betrays little doubt about what happened to the missing lizards. It devotes 270 words to competition from other species, 430 words to disease and parasites, 470 words to wind and solar development, 700 words to off-highway vehicle use, and 1,200 words to agriculture and grazing in the area. The threat from oil and gas development, however, merits 2,600 words. The report also targets exposure to pollutants, global warming, and changes in the lizards' habitat, which are all presumably the fault of the oil and gas industry.

"[I]ncreased oil and gas development in the range of the dunes sagebrush lizard, including seismic exploration, has caused direct and indirect effects to dunes sagebrush lizard habitat," the FWS alleges. "Oil and gas extraction activities have destroyed and fragmented dunes sagebrush lizard habitat and have resulted in population losses...."[5] As evidence, the FWS cites the junk study based on the graduate students' desert strolls. The proposal deemed the wells, along with the roads, trucks, power grids, and pipelines associated with them, to be grave threats to the species' continued existence. Furthermore, the listing proposal alleged that geologists' seismic exploration for oil posed an "imminent threat" to the three-inch lizards.

The Fish and Wildlife Service's proposal to add the lizard to the list of endangered species was a shot across the bow of the oil and gas industry, noting ominously that more than 50 percent of all the oil production in Texas occurred within the habitat of the sand dune lizard. The FWS extracted "voluntary" conservation agreements from the oil producers that gave the agency far-reaching powers in the areas occupied by the lizard—and even in areas that were not occupied but that the FWS deemed "suitable" potential habitats.

In other words, the federal bureaucracy, citing a bogus threat to a hitherto unknown and unremarkable lizard, claimed sweeping authority over the majority of oil production in Texas. The FWS conservation agreements gave bureaucrats the power to direct the routes and construction of new roads, pipelines, and power lines and to limit seismic exploration of potential resources.

The threat of having their operations shut down beat the oil producers into submission. As Ben Shepperd, president of the Permian Basin Petroleum Association, put it at the time, the federal government's position amounted to "We will not list the lizard as endangered if you will cede private and state lands to federal control."[6]

In June 2012 the FWS announced it would drop its proposal to list the dunes sagebrush lizard as endangered based on the "unprecedented commitments to voluntary conservation agreements" it had won from the oil and gas producers.[7] While that may sound like a good outcome, in fact it proves the FWS had abandoned any pretext of scientific credibility for the endangered species list. The agency used its classification authority as a tool for bureaucratic extortion. Why should energy developers accept "conservation agreements" for a species that no evidence shows is threatened? Either the science shows the lizard is endangered, and the bureaucrats must list it, or the science does not show it is endangered, and they must leave private companies alone.

Although the Fish and Wildlife Service has abandoned its campaign to "save" the dunes sagebrush lizard, it hasn't quit trying to seize control of state and private land. Now it proposes to list the lesser prairie chicken as a threatened species on similar grounds. The birds inhabit large swaths of Texas, New Mexico, Oklahoma, and Colorado, but the FWS argues that the birds avoid the "artificial vertical structures [that] are appearing in landscapes used by lesser prairie chickens."[8] The proposal is full of such innuendo about the energy industry's effects on the birds, and a "Lesser Prairie Chicken Interstate Working Group" is developing a conservation plan that oil and gas developers could be pushed to adopt.[9] No one seems to ask how threatened a chicken that spans at least four states can be.

These stunts by the Fish and Wildlife Service are outrageous because their purpose is not to protect wildlife (no evidence suggests the animals needed protection) but to harass energy developers and disrupt the production of oil. The lizard and the prairie chicken are merely the bureaucratic prison guards' pretext for attacking an industry they don't like. This abuse of power corrupts an honorable function of government, the conservation of genuinely endangered species. Sadly, it is not an anomaly in the federal bureaucracy.

In January 2012 the Justice Department filed criminal charges against a number of oil and gas producers in North Dakota for alleged violations of the Migratory Bird Act, originally implemented in 1918. The U.S. attorney for the District of North Dakota, Timothy Purdon, charged the companies in federal court with "taking" several migratory birds found dead on property near their reserve pits. The criminal charges carried fines and, potentially, prison terms.

Continental Resources was charged with "taking" a single Say's phoebe, a bird the size of a sparrow. Brigham Oil and Gas was charged with taking two mallard ducks. Newfield Production Company was charged with taking two mallards, one red-necked duck, and one northern pintail duck. None of these species is endangered.[10]

Were the oil and gas developers guilty of a crime because a few birds found their way into the oil pits? Were the deaths so egregious that the

Justice Department had no choice but to stand up for avian welfare? Apparently the administration thought so.

Yet as the presiding judge noted in his order dismissing the case, the department evidently felt no such compulsion to crack down on the wind-energy industry, which kills thirty-three thousand birds in its turbines each year. Nor, as the judge pointed out, does the department consider it criminal to erect a building with windows (which kill at least ninety-seven million birds a year), to drive a car (sixty million birds a year), to keep a pet cat (thirty-nine million birds), to operate cell towers (five million), or to fly an airplane (tens of thousands).[11] The Justice Department was obviously interested less in defending the Say's phoebes than in persecuting an industry the administration doesn't like. Justice—to say nothing of progress—took a backseat to the prison guards' ideology.

The administrator of the Environmental Protection Agency's Region 6 (Texas and four adjoining states) was forced to resign in 2012 when his candid description of prison guard tactics was made public: "[T]he Romans used to conquer little villages in the Mediterranean. They'd go into a little Turkish town somewhere, they'd find the first five guys they saw, and they'd crucify them. And then, you know, that town was really easy to manage for the next few years.... [T]hat's our general philosophy."[12]

Bureaucratic despotism—just the model to keep America the world's most dynamic economy on top in the twenty-first century! It's not exactly the attitude that the Americans rebelling against King George III intended to instill in a free government of free citizens.

Emmy Award–Winning Lies

The HBO documentary *Gasland* aired in 2010 with great fanfare. It purports to reveal the environmental havoc that fracking is wreaking on small, hardworking towns across the country. Throughout rural America, the film suggests, evil corporations are injecting poison into the ground, cracking open underground aquifers, and destroying drinking water and surface streams for miles around.

Gasland confirms what lots of cultural liberals on the coasts were already inclined to believe about this fracking they'd heard about. It sounds dangerous, and everyone knows that energy companies are up to no good. In any event, it produces more fossil fuels, which means more CO_2 and therefore more global warming. Every polite person opposes it.

"With its jolting images of flammable tap water and chemically burned pets, New York theater-director-turned-documentarian Josh Fox's Sundance-feted shocker makes an irrefutable case against U.S. corporate 'fracking,'" raved the *Village Voice*.[13] The film won the Sundance Film Festival's special jury prize and an Emmy for its director. It was nominated for three more Emmys and an Academy Award.

The most dramatic scene in the documentary depicts the kitchen of a Colorado man who lives in an area dotted by gas wells. Over his sink is a handwritten sign warning guests, "Do Not Drink This Water." As cameras roll, the man approaches his faucet with a lighter and turns on the tap. The water flowing out explodes within seconds, covering his sink in flames. "I smell hair!" the man says as he withdraws his singed arm. "That one was kinda spooky. That even surprised me, and I've been lightin' that water quite a bit."

Practically every review of the movie mentioned this scene, usually including photographs of the flaming tap water or a short video clip of the explosion. The *New York Times* review was headlined "The Costs of Natural Gas, Including Flaming Water," and the online edition included an excerpt of the scene, with a caption explaining that the film "explores the effects of a type of natural gas drilling on residential water supplies."[14] The Huffington Post included the video, too, under the headline "'Gasland' Documentary Shows Water That Burns, Toxic Effects of Natural Gas Drilling" and claimed the faucet clip "shows tap water contaminated with combustible gases from nearby natural gas wells."[15] The scene was so widely distributed and the claim so often repeated that it was probably the first thing many Americans ever heard about fracking.

There is just one problem—the burning tap water had nothing to do with fracking. The investigative journalist Phelim McAleer looked into the claims and discovered that people in the area had known for decades that natural gas contaminated their water, long before fracking began in the region. The methane deposits just below the surface were a common natural phenomenon there. (McAleer later made a film of his own, *FrackNation*, rebutting *Gasland*.) When he confronted Josh Fox, the filmmaker, Fox conceded he was aware of reports dating back to 1936 of people who could set their water on fire.[16] Even with the methane, moreover, the water is perfectly safe for human beings to drink. But the film discloses none of this information, and in fact blames fracking for the contamination. *Gasland*'s punch line, the scene millions of people read about, is a lie.

Energy in Depth, the research and education arm of the Independent Petroleum Association of America, has revealed several other deceptions in *Gasland*. Dead fish shown floating in a Pennsylvania creek, supposedly casualties of fracking, were killed by an algae bloom. The documentary asserts that fracking caused residents of a Texas town to display increased benzene toxins in their blood; in fact, the state had investigated and found normal levels except in smokers, which was expected. *Gasland* blames fracking for a decline in populations of three "endangered" species. Yet none of the species is endangered, and all are doing fine.[17]

Even though much of *Gasland* is simply misinformation, the journalists and reviewers who wrote glowing reviews of the film's shocking scenes never bothered to check up on the claims themselves. The truth about the methane in Colorado water supplies, for instance, would have been easy to discover. When McAleer, an independent journalist, finally forced the facts into the open, the media's coverage of the errors did not match the intensity of their original coverage of the documentary. Indeed, many outlets didn't follow up at all. To this day, many people believe that fracking produces flammable drinking water.

One of the radical environmentalists who colluded with Josh Fox in *Gasland* was Alfredo Armendariz, then a teacher at Southern Methodist University. Shortly after his appearance in the film, he was appointed regional administrator of the EPA in Texas—the same official who would later resign after comparing the EPA's enforcement techniques to "crucifixion."[18]

The EPA itself has been trying for years to find an excuse to restrict fracking, and on a number of occasions it has come close. In Texas, the agency investigated a case of alleged contamination for more than a year before conceding there was no evidence fracking was to blame. The public face of that anti-fracking campaign, a local homeowner, had produced a widely disseminated video showing flames coming out of his garden hose. The man's water well, it turned out, had been drilled straight into a naturally occurring gas deposit just below the surface.[19] A judge found there were grounds to believe he had, "under the advice or direction" of a local environmental activist, chosen to "intentionally attach a garden hose to a gas vent—not to a water line—and then light and burn the gas from the end nozzle of the hose." The stunt, the judge theorized, "was done not for scientific study but to provide local and national news media a deceptive video, calculated to alarm the public into believing the water was burning."[20] Nonetheless the scene was included in *Gasland II* this year.

Complaints that fracking had poisoned the wells of eleven families in Dimock, Pennsylvania, inspired a major Hollywood motion picture, *Promised Land*, starring Matt Damon and John Krasinski. The film's anti-fracking activist hero turns out to be a double agent employed by an evil energy corporation to discredit the environmental movement. Curiously enough, *Promised Land* "was financed in part by a company that is wholly-owned by the government of the United Arab Emirates."[21] The royal family of the UAE, like the environmentalist Left, isn't eager to see fracking succeed in the United States—but for very different reasons. They're happy to fund America's green prison guards if it means we keep buying oil from the Middle East.[22] In Dimock, too, the EPA and the state environmental protection department found no evidence of serious contamination and no contamination from fracking.

Perhaps the most high-profile claims of groundwater pollution were in Pavillion, Wyoming, population 231. The EPA's troubled four-year investigation of the wells there made national headlines, especially when the agency released its draft report in 2011 theorizing, for the first time anywhere, an "impact to ground water that can be explained by hydraulic fracturing."[23] The major news outlets, already primed with propaganda from environmentalists, ran with the story. The Associated Press trumpeted, "EPA Implicates Hydraulic Fracturing in Groundwater Pollution at Wyoming Gas Field."[24] "E.P.A. Links Tainted Water in Wyoming to Hydraulic Fracturing for Natural Gas," the *New York Times* reported enthusiastically.[25]

In fact, the EPA had conducted its test outside of town near a natural gas reservoir. The USGS had noted naturally occurring chemicals in the town's well water for several decades. Moreover, as Energy in Depth reported, the EPA's own drilling process for its test wells could have contaminated the results.[26]

You can imagine the prison guards' surprise in June 2013 when the agency, after redoing its tests, quietly dropped the case without issuing a final report, "abandoning its longstanding plan to have independent scientists confirm or cast doubt on its finding that hydraulic fracturing may be linked to groundwater pollution in central Wyoming."[27]

To date, there is not a single confirmed instance of groundwater contamination resulting from hydraulic fracturing, after hundreds of thousands of wells have been fracked.

Facts Matter

Americans care about the environment. We want clean air and clean water. We want to protect biological diversity and to conserve our country's natural beauty for future generations. If fracking really is risky, we want to know. But the debate must be based on the facts, and the green prison guards have no interest in a debate like that. Like the teachers' unions responding to innovations in education, the green prison guards are not interested in making the technology of energy production better or safer. They simply want to kill it. The environmental Left has an aesthetic, not a scientific, opposition to fossil fuel, and the facts don't matter. They hate it and the industry that produces it.

The Fish and Wildlife Service, the Department of Justice, the EPA, and environmentalists have all launched attacks on energy producers that were intended not to ensure the safe production of oil and gas but to restrict and destroy the industry.

Greenbacks for the Greens

Not every energy company gets the "restrict and destroy" treatment. Some get the "coddle, fund, and protect" approach.

In his weekly radio address of July 3, 2010, President Obama had words of praise for a Colorado solar panel manufacturer, Abound Solar. The company, he boasted, "will manufacture advanced solar panels at two new plants, creating more than 2,000 construction jobs and 1,500 permanent jobs."

"A Colorado plant is already underway," the president continued, "and an Indiana plant will be built in what's now an empty Chrysler factory. When fully operational, these plants will produce millions of state-of-the-art solar panels each year."[28]

The president had reason to root for Abound apart from the imagined jobs for Colorado and Indiana workers. The company was one of the administration's marquee recipients of stimulus money under a Department of Energy loan program for green tech. Interior Secretary Ken Salazar had already visited an Abound facility the previous year. Companies like Abound would produce the "green jobs" that the White House insisted would drive the economic recovery.

The icing on the cake of the Abound deal was that one of its largest investors, the billionaire Pat Stryker, was a major Democratic donor. She gave $50,000 to President Obama's first inauguration and bundled $87,500 more. She also donated $35,800 to the president's reelection effort and spent at least $220,000 on other Democrat-affiliated groups in 2012.[29] According to the visitor log, Stryker came to the White House in October 2009, before President Obama announced that Abound would receive $400 million in DOE loan guarantees.[30] The White House has not disclosed details about the meeting.

Stryker's relationship with the Democrats was not unusual for a recipient of a Department of Energy loan guarantee. Prologis received a $1.4 billion loan guarantee; the brother of company board member Lydia Kennard bundled at least half a million dollars for Obama in 2008.[31] NRG Energy received $3.5 billion in loan guarantees; Arvia Few, wife of NRG executive Jason Few, raised between $50,000 and $100,000 for

Obama's reelection.[32] And, famously, Solyndra executives donated more than $100,000 to the president.[33]

There was, however, one big problem with Abound Solar. Despite the president's praise, its solar panels were duds. As a source inside the company told the Daily Caller, "Our solar modules worked as long as you didn't put them in the sun." Apparently, sources said, "the company knew its panels were faulty prior to obtaining taxpayer dollars... but kept pushing product out the door in order to meet Department of Energy goals required for their four-hundred-million-dollar loan guarantee."[34]

The ratings agency Fitch gave Abound a "B" credit rating ("highly speculative") in November 2010, projecting a 45-percent chance of recovery. The loan was approved the next month.[35] In July 2012 the company filed for bankruptcy and within months was under investigation for securities fraud.[36] It is left with thousands of "unsellable" solar panels to dispose of.[37] Taxpayers will lose tens of millions of dollars.

Why do green energy companies get multi-billion-dollar loan guarantees or millions in grants from the government while other energy companies are attacked by the federal bureaucracy at every turn? Because our leaders are committed to forcing an outdated vision of the future on the American people. The exotic replacements the administration advocates were conceived as solutions to the mythical problem of "peak oil." These were always expensive alternatives, but when many Americans believed the world was running out of oil, some thought it was pragmatic to develop them.

We know for certain, however, that the world is not running out of oil—in fact, we have more than ever, and the estimates keep going up. Yet the environmental Left's hostility to the fuel our economy depends on, and their dedication to forcing a transition to boutique alternatives, remains undeterred. They are like devotees of vinyl records or Betamax videocassettes, refusing to give up even though times have changed. There is one important difference, however: the green prison guards have the coercive power of government on their side, and they're employing it aggressively.

President Obama is not the only one pushing the green-energy vision, but he certainly jumped to the head of the parade. In his victory speech after the 2008 Iowa caucuses, he promised to "free this nation from the tyranny of oil once and for all."[38] Then he went to the automotive capital of the United States and compared "the tyranny of oil" to "fascism and communism."[39] Even as estimates of American oil resources were skyrocketing, he vowed to "end the age of oil in our time."[40]

As president, Obama has assumed the role of green prison guard in chief, committed to forcing an expensive and now unnecessary transition to exotic replacements. Harold Hamm, one of the pioneers of the Bakken oil field, had a revealing conversation with the president, which he recounted to the *Wall Street Journal*'s Stephen Moore:

> "I told him of the revolution in the oil and gas industry and how we have the capacity to produce enough oil to enable America to replace OPEC. I wanted to make sure he knew about this."
>
> The president's reaction? "He turned to me and said, 'Oil and gas will be important for the next few years. But we need to go on to green and alternative energy. [Energy] Secretary [Steven] Chu has assured me that within five years, we can have a battery developed that will make a car with the equivalent of 130 miles per gallon.'" Mr. Hamm holds his head in his hands and says, "Even if you believed that, why would you want to stop oil and gas development? It was pretty disappointing."[41]

Green energy was a headline feature of the stimulus, a ninety-billion-dollar "investment" that would "save" Americans money.[42] The president was apparently taken in by the "green jobs" fad, believing solar panels and wind turbines would drive the economic recovery. In his book *The Escape Artists*, Noam Scheiber reveals Obama's "particular obsession" with green energy; the president asked his advisors "week after week" where all the green jobs were.[43]

Since there was no green energy sector to provide the millions of green jobs they promised, the prison guards set out to create one. They engineered government programs to support handpicked technologies at every step of the production process. And they embarked on this quixotic project at the expense of American taxpayers in the middle of the worst economy since the Great Depression.

If you wanted to launch a green-energy project, $12 billion in grants was available from the Treasury and Energy Departments. You didn't even have to give the taxpayers—your "investors"—a stake in the company.[44]

If you needed a commercial loan for your green-tech startup, the government would underwrite it. Obama's stimulus legislation contained $14.5 billion in loan guarantees from the Department of Energy.[45] Of the twenty-six recipients of those guarantees, twenty-two were not even creditworthy, receiving junk ratings.[46] Sixteen loan guarantees went to ventures in solar energy, a technology unlikely to be commercially viable after the fracking-induced collapse in natural gas prices.[47] This was just one of six federal loan programs for green energy.[48]

Still need special tax advantages to keep your green business open? You can get tax credits for producing green electricity,[49] for making capital investments in green energy,[50] for conserving energy,[51] and for building energy-efficient buildings.[52]

Need federal dollars because nobody wants to pay for your green upgrades? You can get up to $3 million in federal grants for "improvements" in rural energy installations.[53] You can get up to 50 percent of total project costs for converting "biorefineries" to operate with "renewable biomass."[54] Rural agricultural producers and small businesses can get up to 25 percent of the cost of renewable energy systems to help them "become energy efficient" and "use renewable energy technologies and resources."[55]

If you are in the solar energy business and foreign products are cheaper than yours even after all the federal subsidies for you and your customers, the government will give you a boost by hitting the Chinese competition with a tariff of between 24 and 36 percent.[56] If you make wind turbines, the government adds a tariff of up to 72 percent on your rivals' products.[57] (How driving up the cost of solar and wind energy equipment advances the wide adoption of green tech is unclear.)

Do you want to export green products abroad? The Export-Import Bank is required to "allocate 10 percent of its annual financing to renewable energy and environmentally beneficial products and services."[58]

Does your green company suffer from a shortage of green workers? For this unlikely situation, the stimulus allocated $500 million for training programs to "prepare workers for careers in energy efficiency and renewable energy."[59]

Do you need customers for your green tech? Every federal department, including the Department of Defense, is ready to invest in "sustainability."

Is there no demand for the overpriced power your boutique energy firm generated? Government will simply force consumers to buy it. In thirty-seven states, renewable portfolio standards require electric utilities to get a certain portion of their power from renewable sources. Electric utilities in California, for example, must get 33 percent of the electricity they sell from green sources by 2020.[60] The explicit purpose of these standards is artificially to support green-energy providers. Home electricity costs in these states are on average 32 percent more expensive than in states without such mandates.[61]

Do you make a green consumer product and need a little extra help? The federal government can cut your customers some big breaks. They can take tax credits for 30 percent of a variety of renewable energy systems for their homes.[62] The stimulus included credits for homeowners of up to $1,500 for energy efficiency improvements.[63] Rural agricultural producers and small businesses can get grants for up to 25 percent of project costs to purchase renewable energy systems.[64]

In all, the federal government has nearly *seven hundred initiatives* devoted to supporting a green-energy agenda at an enormous cumulative cost.[65] These programs keep Americans trapped in a world of artificial scarcity and expensive energy, and they have failed to produce breakthrough technologies. In fact, the heavy federal involvement in green tech has probably inhibited breakthroughs. By picking certain companies or technologies to be "winners," the government doomed their competitors to be "losers." In a normal market, consumers make those choices. They vote with their wallets. The businesses that offer the best value rise to the top. But when the government gives advantages to some companies or technologies not because of the promise of the product but, for example, because of whom the investors know, it distorts this natural process. Bad ideas live on, while good ideas, which can't compete with their privileged and protected competitors, fail.

Nothing illustrates the catastrophic results of picking winners and losers as well as the Department of Energy's loan guarantee program, the centerpiece of Obama's stimulus legislation. The program selected some green companies for federal underwriting on loans worth hundreds of millions, sometimes billions, of dollars. How can anyone compete with that?

A remarkable number of the businesses that received loans were, like Pat Stryker's Abound, friends of the Obama administration, and the White House was intensely involved in the process of selecting them.

At least some senior officials were aware of the program's serious flaws. Jim McCrea, who was in charge of credit analysis for the loans, wrote in an email to colleagues, "I really cannot fathom how one figures out whether a loan to a PV [photovoltaic] manufacturer is being made to one that will survive. Everything about the business argues for the failure of many if not most of the suppliers.... All in all in the solar field, *I think it is extremely easy to pick losers and I really do not know how to pick winners.*"[66]

This is a man whose job is to pick winners in the solar field!

Lawrence Summers, one of President Obama's chief economic advisors, apparently agreed with McCrea's observation. A venture capitalist who was invested in Solyndra wrote to Summers in December 2009, warning, "While that [loan] is good for us, I can't imagine it's a good way for the government to use taxpayer money." As Reuters reported, "Summers agreed that the government is a 'crappy VC' (venture capitalist)."[67]

The government provided assistance to companies that clearly did not need it, harming their competitors while unjustly enriching private investors. In a memo to President Obama, three of his top economic advisors (including Summers) offered Shepherds Flat as an example of a project that "would likely move forward without the loan guarantee."[68] Another company, NRG Energy, received $3.5 billion from taxpayers despite generating more than $9 billion in revenue in 2011.[69] It is a Fortune 500 company. Cogentrix received more than $90 million despite being wholly owned by Goldman Sachs.[70] Prologis received a $1.4 billion loan guarantee despite revenues of nearly $2 billion in 2011.[71] And Ormat Technologies received $350 million despite revenues approaching $450 million in 2011.[72]

It is little wonder that an industry in which the government arbitrarily determines which companies have a chance and which ones don't has seen so little of the progress President Obama promised. If the cell phone market operated by those rules, we would never have gotten the iPhone. Indeed, under those circumstances, Steve Jobs would never have bothered to create it. Green tech is no different. The green prison guards are not just holding our oil and gas resources prisoner but their pet industry as well.

A Convenient Excuse

The only way to make solar and other green energy sources economically competitive is to drive up the cost of the abundant and affordable alternatives—oil and gas. Why would anyone pursue policies that amount to economic sabotage? Because, the environmental Left believes, a shift from fossil fuels is necessary to prevent impending disaster from climate change.

The fears that drive this perverse agenda can be fairly described as apocalyptic. A 2013 article in *Rolling Stone* titled "Goodbye Miami: Why the City of Miami is Doomed to Drown" is representative of what keeps these zealots up at night. From its imaginative perspective of the end of the twenty-first century, the article envisions a parade of catastrophes due to global warming, from highways that "disappeared into the Atlantic" to the dumping of "hundreds of millions of gallons of raw sewage into Biscayne Bay" to storm waters that "took weeks...to recede." The metropolis we know "became something else entirely: a popular snorkeling spot where people could swim with sharks and sea turtles and explore the wreckage of a great American city."[73]

Rolling Stone was far from the first to promote this kind of alarmism. Al Gore's documentary *An Inconvenient Truth* features maps depicting the large swaths of the world that will be flooded if we don't immediately stop using the fuels on which our economy depends.

Proponents of the climate change theory might well be correct that temperatures will increase and sea levels will rise in the future. It is, however, important to be clear about a few points. First, all of the catastrophes the global warming alarmists prophesy are based on computer models that are supposed to simulate how weather works. The predictions of warmer temperatures, the rise in ocean levels, the frequency of extreme weather—they're all based on computer programs that forecast the weather fifty or even a hundred years from now. But of course, we have no such precise understanding of how the climate works. We know that weather is a mind-bogglingly complex phenomenon, so much so that no scientist would claim to know with any precision what it will be a month from now. (Such a forecast, after all, would be quickly disprovable.) Yet when the time horizon is a few decades out, they are somehow more comfortable with their forecasts.

Climatologists' computer models have a long history of being spectacularly wrong. In his book *Chaos: Making a New Science*, James Gleick recalls, "The fifties and sixties were years of unreal optimism about weather forecasting. Newspapers and magazines were filled with hope for weather science, not just for prediction but for modification and control.... There was an idea that human society would free itself from weather's turmoil and become its master instead of its victim."[74] John von Neumann, the father of modern computing and one of the greatest minds of the twentieth century, promoted this hope in frequent lectures about the promise of computers. With sufficiently powerful machines, Gleick writes, "Von Neumann imagined that scientists would calculate the equations of fluid motion for the next few days. Then a central committee of meteorologists would send up airplanes to lay down smoke screens or seed clouds to push the weather into the desired mode."[75]

In reality, even the most advanced computer models fall far short of such precision. The reason, Gleick observes, is that "for small pieces of weather—and to a global forecaster, small can mean thunderstorms and blizzards—any prediction deteriorates rapidly. Errors and uncertainties multiply, cascading upward through a chain of turbulent features, from dust devils and squalls up to continent-size eddies that only satellites can see."[76] Weather forecasters cannot predict the temperature even a few weeks in advance. Yet climate-change theorists are predicting aggregate temperatures to a tenth of a degree at the end of the twenty-first century.[77]

Global-warming alarmists forget that their models are human creations, vague abstractions trying to predict events in the real world. They are not *facts*. This was a lesson I learned early in my career. In the 1970s, when I first taught environmental studies, we used the popular book *The Limits to Growth*. It related how a group called the Club of Rome had developed a computer program and fed all sorts of data about economic growth and natural resources into it, producing predictions of disastrous shortages. There is, however, a principle in computing known as GIGO—"garbage in, garbage out." Every prediction of *The Limits to Growth* eventually proved wrong. Every projected shortage was averted. The price signals of a free economy worked. New technology and entrepreneurial creativity created solutions faster than the problems developed.

Another prophet of doom, Paul Ehrlich, achieved fame at about the same time with *The Population Bomb*, a collection of some of the most stupendous miscalculations ever published.

None of this is to say that the scientists who predict global warming are ignorant or malicious. No doubt they craft the best models they can. But the limits to our ability to understand infinitely complex systems demand a big dose of humility. Indeed, the impermissibility in polite company of questioning the computer models is reason enough to be skeptical. Such hubris is antithetical to the scientific method.

Adapting to Change

Another important point is too often overlooked in the climate change debate. Climatologists are climatologists. They are not political scientists, economists, or engineers. So while they *might* be able to tell us what the weather will be, they are not the people to tell us what we must do about it. Yet the discussion usually jumps directly from what the "science" says to why we should accept the fantastic remedies prescribed by the environmental Left—which always involve impoverishing the country and increasing government power.

The assumption is that the only reasonable response to an increase in temperature of a couple degrees and a rise in sea level of a few inches is to *prevent* them from happening—to control the climate and the seas. There is an alternative, however. It's how human beings have responded to weather and the oceans for hundreds of thousands of years: adapt to it.

When you think of the Netherlands, you probably envision sweeping landscapes dotted with windmills. The Dutch erected thousands of them in the sixteenth and seventeenth centuries. Perhaps you have assumed they built them to grind wheat. Of course, the windmills had agricultural uses, but their main purpose was to pump water—not for drinking or irrigation, but to keep the country dry. About a third of the Netherlands lies below sea level. Without human intervention, much of the country would be underwater, including the entire city of Amsterdam.

The Dutch ingeniously constructed a system of dikes, dams, canals, and drainage ditches to vacate the ocean from their country's broad plains. The windmills pumped water from low-lying fields into drainage canals. They were the breakout technology of their day.

Much of the country would be under water today if not for the complex system that walls off the sea. Over a period of nearly eighty years starting in 1920, the Netherlands constructed a system of twenty-two giant dams, dikes, and flood barriers. Rotterdam lies twenty feet below sea level, walled off from the water by a pair of massive swinging gates.[78] The same basic technique has kept the country dry for more than four hundred years (although there have been a few severe floods). In all those centuries, the Dutch never considered lowering the sea.

Many of the most densely populated places on earth have taken a cue from the Netherlands to expand their available real estate. In 2007, construction crews on the site of the World Trade Center in Manhattan discovered the remains of an eighteenth-century ship. Archeologists determined that it had been hoisted onto the docks that once marked the edge of the island and abandoned there. Soon after, in the 1790s, New Yorkers began extending the city into the Hudson River. "By the 1830s," *Archaeology* reported, "the area had been completely filled and the new shoreline lay 200 yards west of its original location at modern Greenwich Street. Over the decades, the earth brought in for the shoreline extension—and the trash discarded there—completely covered up the ship."[79]

Indeed, humans have transformed the island of Manhattan dramatically since Henry Hudson discovered it in 1609. The skyscrapers that house the offices of Standard & Poor's, Morgan Stanley Smith Barney, and Bank of New York Mellon all stand where the river once flowed. Further uptown, what is now Stuyvesant Town and Peter Cooper Village (neighborhoods composed of fifty-six high-rise apartment buildings) would have been sitting in fifteen feet of water in 1609. A huge swath of Chelsea, too—almost everything west of where the High Line is today—would have been under water. The major thoroughfares that run the length of the city, FDR Drive, the West Side Highway, and the Henry Hudson Parkway, would have stood some distance out into the river. Even much of the inland part of the island was a tidal estuary.[80] Lower Manhattan's Battery Park City, about ninety-two acres of land, sprang out of the water as recently as the 1970s.

In Boston, the story is even more dramatic. Most of what is now Boston was under water when the first English settlers arrived in 1630. The Boston Garden, Fenway Park, Faneuil Hall and Quincy Market, the Prudential Building, and Copley Square would all have been under water in the Puritans' Boston. In fact, the city's residents have been pushing back the sea almost since the town was founded. A massive landfill project beginning in the early 1800s cut the top off of Beacon Hill to begin filling in the Back Bay. Over the century, Bostonians continued to bring in earth to extend the shoreline, dramatically altering the city's shape by 1900. The Boston area's natural contours are completely unrecognizable on a modern map.[81]

It's impossible to know whether the computer models' projections of slightly warmer temperatures and slightly higher seas will turn out to be accurate over the course of our century. But it is easy to see that many of the worst catastrophes the global-warming alarmists forecast—such as Miami's or Manhattan's turning into Atlantis—will never come to pass. That's because even if the models prove correct, human beings won't sit idle while the sea level climbs a few inches. We'll adapt, just as we have for centuries.

Anyone who thinks the next breakout is inevitable should consider the fanatical determination of today's green prison guards, who would impose trillions of dollars in taxes and regulations in their quest to keep the planet's climate from changing.

CHAPTER SIX

BREAKOUT IN TRANSPORTATION

The Pain of Gridlock

The unbearable commutes that many of us endure ten times a week are powerful testimony to the American work ethic. Given the traffic congestion that brings many of the nation's cities to a halt twice a day, it's remarkable that America's drivers don't snap.

One of the senior members of our staff recently moved with his family to Richmond, Virginia, from a home near our office just outside Washington, D.C. Several days a week, he now makes the trip to Washington for work. Without traffic, his drive of about eighty miles would take an hour and a half on I-95. But the highway from Richmond to Washington is never without traffic. As our last meeting of the day wraps up and he checks Google Maps on his iPhone, the line tracing his commute frequently has turned deep red (indicating standstill traffic), and the estimated trip home is well over four hours. Anyone who has ever been to Washington knows that the city saves its worst gridlock not for Congress but for the city's drivers.

The gridlock epidemic extends far beyond our nation's capital. It clogs cities from coast to coast. A CBS News report last year quoted an Atlanta man who described a daily commute that would probably be illegal if inflicted on Guantanamo detainees: "I live about thirty-two miles west of my office," he said. "The entire drive takes anywhere from one to two hours each way. What makes it brutal is that the road seems to drive directly at the sun. If you are not prepared, this sun can be literally blinding. When cars turn towards the sun, traffic comes to a screeching halt. Many are unprepared, which leads to wrecks and slowdowns. Going home, the entire process is repeated as the sun going down makes traffic a nightmare."[1]

Millions of Americans endure these monstrous commutes every day. The average commute in San Francisco is thirty-two minutes each way. In Boston, it's thirty-three minutes. In New York, thirty-nine minutes.[2] For 3.2 million of us, the drive to work is more than ninety minutes. And 16.5 million drivers, or 12.5 percent of all commuters, leave home before 6:00 a.m.[3] Much of this time is spent at a standstill or alternately accelerating and slamming on our brakes, coffee sloshing back and forth in its cup holder. In 2009 the average American spent thirty-four hours sitting in traffic congestion. That adds up to $87.2 billion in wasted fuel and lost productivity—$750 per traveler. In big cities with large commuting populations, it was even worse: seventy hours a year lost to congestion in Los Angeles, sixty-two in Washington, fifty-seven in Atlanta.[4] Breaking out of the gridlock prison would be an enormous improvement to the quality of life of millions.

The hours behind the wheel can be more than an inconvenience, as well. Each year, forty thousand Americans die in car accidents, and two million are injured—a number equivalent to the population of Houston.[5] Car accidents are the leading cause of death for Americans under thirty-five. There's a good chance that driving a car is the most dangerous thing you do regularly, even if you don't think about it.

Ending those traffic deaths—not to mention traffic gridlock—would be almost as big a breakthrough as curing breast cancer or preventing heart attacks. But it would mean making the biggest changes to the automobile since the Model T Ford.

The Self-Driven Future

When we left Sebastian Thrun, the Google vice president and founder of Udacity, he was working to undermine the education establishment and bring a high-quality college degree to the average American for less than 10 percent of the current cost. But you may recall that Udacity is only the latest of Thrun's potentially world-changing projects. The first Stanford course he offered online, which attracted 160,000 participants, was on artificial intelligence. Thrun has spent much of his career on AI and machine learning, and he began working on a variety of self-driving automobiles for competitions run by the Defense Advanced Research Projects Agency (DARPA) after he arrived at Stanford in 2003.

The DARPA Grand Challenge, as the agency called it, sought to jump-start the development of genuinely autonomous vehicles in the hope of converting a large portion of military ground forces to drone vehicles. As the Defense Department was reminded during two ground wars in the Middle East, resupply convoys, emergency medical evacuations, and intelligence or forward observation missions can be a dangerous business. Indeed, many of the casualties in Iraq and Afghanistan have been inflicted on supply convoys traveling through hostile territory. DARPA is eager to take human beings out of these perilous tasks.

When Thrun began working on Stanford's Grand Challenge entry in 2004, DARPA had already held the first off-road race through the barren, sometimes mountainous landscape of the Mojave Desert. No team had managed to make it more than seven and a half miles down the 150-mile course. The 2005 race followed a similar route, and the Stanford team, led by Thrun, entered a modified cobalt blue Volkswagen Touareg R5 that they called Stanley (not exactly a Humvee but true to Thrun's German roots). The team equipped Stanley with GPS, a camera, five laser rangefinders, and a radar system, all of which fed data into six computers in the trunk. These machines, running custom software, drove the car.[6]

Although it was up against sturdier off-road vehicles, including a number of Humvees and a sixteen-ton, six-wheeled Oshkosh truck, Stanley was the front-runner almost from the beginning of the race. DARPA staggered the start times in order to give each vehicle some space, but little more than halfway through, the organizers had to push the pause button on Stanley for several minutes (briefly suspending both the SUV and the clock) because it was getting too close to Carnegie Mellon's entry, a bright red Humvee called H1ghlander. Less than six minutes after Stanley resumed the race, it caught up with H1ghlander again, and again DARPA officials paused the vehicle—this time for even longer. Eventually, five and a half hours into the race, DARPA paused H1ghlander so Stanley could pass. The Stanford team won first place, finishing in just under seven hours.[7] Four other teams finished as well, including Carnegie Mellon (where incidentally, Thrun had helped lead the Robot Learning Laboratory before joining the faculty at Stanford).

Five vehicles driving autonomously through 130 miles of desert mountains marked a sharp improvement over the first race, in which the competitors barely got out of the gate. But it was still a long way from street-ready. As anyone who has ever driven a car understands, the real challenge is dealing with everybody (and everything) *else* on the road. In the 2005 race, the robots didn't have that problem at all; the DARPA officials had paused them whenever they came close to another vehicle. Nor had they faced any of the other hazards that human drivers routinely navigate, like fallen branches, bicyclers, joggers, or children chasing balls into the road.

In 2007, Thrun and his team at Stanford entered the DARPA Urban Challenge, which took autonomous driving from the desert to a fake town, a much more complex course, complete with traffic, stoplights, traffic signs, and various hazards. For this more forgiving terrain, Thrun and his colleagues used a blue Volkswagen Passat wagon called Junior, which they equipped with fancier sensors capable of composing a detailed model of the car's surroundings (including, now, moving objects). Competitors now had to handle intersections, roundabouts, merges, one-way roads, parking spots, encounters with other autonomous cars (some of which were liable to do strange things), and more. DARPA rules permitted vehicles to stop and analyze a given situation (as a human driver might occasionally do) for no more than ten seconds.[8]

Competing vehicles had to complete a set of tasks within six hours while other competitors were also on the road. DARPA scored the teams not solely on time but also on how well the cars handled the various trials thrown at them. Six teams finished the day, and Stanford's Junior earned second place (this time edged out by Carnegie Mellon). The Passat's average speed was 13.7 mph.[9]

The DARPA challenges had provided proof of concept, showing that autonomous cars could eventually become smart enough to handle many of the complex tasks of driving on civilian roads. But there was still a long way to go between the controlled test course, even of the urban challenge, and the chaotic California roadways.

Thrun met Google cofounder Larry Page through the DARPA challenges, and after the race Page approached him about coming to work for Google on some groundbreaking projects (including Street View). Thrun accepted. "Google is run by exceptional, visionary leaders with very deep technical curiosity and an ambition and a calling that goes way beyond organizing the world's information," Thrun told me of his decision. "It became clear that with the work we did at Stanford University, in my lab...there was an opportunity to fundamentally transform society."

It made sense to go to Google, he said, because the challenge of self-driving cars is "not an automotive problem, it's not an engineering problem in the classical physical, electronic, or mechanical sense. It's really a problem in terms of computer science, and Google is really one of the leaders in the world on that, with a scale of computing and staffing that was hard to get at any university."

Thrun assembled a team of twelve engineers, including some of the people who worked with him on the DARPA challenges.[10] Soon, they were loading up Toyota Priuses with LIDAR sensors and trunks full of computers. They began testing the cars on public streets, with human drivers behind the wheel in case they needed to take control of the car in a dangerous situation. Early on, human intervention was common. "In the beginning," Thrun said, "we were able to drive maybe in the order of ten miles between critical incidents." The more they tested the cars, the more unexpected situations they encountered and programmed the cars to handle properly—when people ran out into the street, for instance, or when a driver in front of them suddenly changed lanes.

In fewer than three years, the team had advanced and tested the technology far more thoroughly than anyone knew. In October 2010, Thrun took to Google's official blog to announce:

> We have developed technology for cars that can drive themselves. Our automated cars, manned by trained operators, just drove from our Mountain View campus to our Santa Monica office and on to Hollywood Boulevard [more than 350 miles]. They've driven down Lombard Street, crossed the Golden Gate bridge, navigated the Pacific Coast Highway, and even made it all the way around Lake Tahoe. All in all, our self-driving cars have logged over 140,000 miles.[11]

That's greater than the lifetime of an average car. By the summer of 2013, the cars had driven themselves more than six hundred thousand miles, virtually without incident. "Our work is focused on building cars that can drive anywhere by themselves, any street in California," Thrun told the TED conference in 2011. Video shows Google's car deftly dodging deer, joggers, tollbooths, people crossing the street, and bad drivers. The car had advanced so far beyond the 13.7 mph average speed of the DARPA Urban Challenge that other drivers hardly took notice of it. And instead of needing human intervention every ten miles, Thrun told me, "now we are at fifty thousand miles, which is the equivalent of four years of human driving."

Still, that's not quite enough for Thrun, who wants to get the technology so perfect that human beings could sleep behind the wheel and cars could drive themselves with nobody inside. "I want to get to a point where we vastly exceed human safety," he said. "To the point where we feel confident that we can drastically reduce traffic fatalities before we launch it."

Now that Google, a search-engine company, has achieved the greatest breakthrough in automobiles since the internal combustion engine, essentially as a side project, the legacy car companies are racing to catch up. Toyota, Nissan, BMW, Audi, and Volvo have all demonstrated prototype self-driving cars of some capacity, although none seems to have matched Google's performance. One thing is certain: now that Thrun and his fellow pioneers have created the technology, Americans are going to want it—from whatever manufacturer can provide it.

★ ★ ★ ★ ★

If we have self-driving cars, how will we use them? If the millions of Americans who spend nearly an hour each day commuting to and from work have anything to say about it, we'll probably want to start by taking a bite out of traffic congestion.

"Congestion" is a relative term. It might be hard to believe, but at the height of rush hour, vehicles occupy only 6 percent of the space on our highways. Think of an egg carton with only one egg in it. But with human drivers, that's as close as cars can safely travel. Each year, highway use increases by about 3 percent, so the roads are nearing their limit, and we're stuck in gridlock.

If cars could be packed more tightly on our roads, Thrun points out, we could double or even triple highway capacity. Fitting three cars into the space that now holds one would be the equivalent of adding four lanes to a two-lane highway. With self-driving cars, we could achieve that density, since they are constantly alert and they can react instantaneously. By the time they reach the market, moreover, they'll probably be able to talk to each other—pinging the cars behind them with notices to slow down or speed up. The SARTRE project in Europe, an effort that includes Volvo and Ricardo UK as partners, has already demonstrated a "convoy" system of car-to-car communication that allows a lead truck to direct a train of cars following closely behind. (The configuration also yields substantial fuel savings—up to 21 percent—because of the decreased wind-resistance.)

Just as self-driving technology would let cars drive closer together, it would also allow them to travel much faster. Greater density at greater speeds could virtually eliminate the problem of traffic, even with no new roads. (The question remains how we can achieve these benefits fully if highways have a mix of traditional cars and self-driving cars. The solution might be roads reserved for self-driving vehicles, just as many states now reserve certain lanes or highways at peak hours for "high-occupancy vehicles.")

Self-driving cars could do something even more important than eliminating traffic congestion—they could be dramatically safer. If they eliminated the overwhelming majority of accidents, we could make them out of lighter materials and substantially improve fuel economy. Self-driving cars could transform the lives of elderly and disabled persons, as well, giving them previously undreamed-of independence. Drunk driving would become a thing of the past.

Self-driving technology would revolutionize the shipping industry—businesses like UPS and FedEx, as well as long-haul trucking. Many shipping routes would no longer require drivers, and they could operate around the clock.

In 2013 I visited Caterpillar, which has been working with the Robotics Institute at Carnegie Mellon University on a self-driving truck for mining and construction sites. These are huge industrial devices, two stories tall and carrying loads of four hundred to five hundred tons. In an open-pit or strip mine, a self-driving truck could work seven days a week, twenty-four hours a day apart from time for routine maintenance. In some remote, high-cost areas, a self-driving truck could save up to a million dollars a year. Caterpillar is not alone in the quest for self-driving trucks. Rio Tinto is buying 150 self-driving trucks from Komatsu. Oshkosh Defense is building the TerraMax self-driving truck for the Defense Department. These developments are happening now.

The taxi business, and any business that relies on professional drivers, will face major disruptions. (This is one more example of why we will need a 24/7 online adult learning system for people displaced by technological change to learn new skills and new capabilities.)

It might be less obvious that adding self-driving capability to cars could similarly disrupt the automotive manufacturing industry. But the big automakers could be looking at a fundamental challenge to their business model. Today, the average American family owns two, maybe even three, cars. If your family is typical, every one of your cars sits idle in the driveway or in a parking lot twenty-two hours or more a day. That's a lot of downtime for some very expensive equipment, but it's often necessary because mom needs to drive to work, dad needs to drive to work, and the teenage kids need to drive to school and their babysitting jobs.

When families have genuinely self-driving cars, however, they may find that one is enough for the whole family. It ferries the kids to school at 7:30 while mom and dad get ready for work. It returns home by 8:00, departs with mom and dad at 8:30, drops dad off at work by 8:45, and mom by 9:00. If dad has a lunch meeting downtown, the car is waiting for him by noon, ready to shuttle him into the city. The commute home is fast and leisurely, and both parents are enjoying a quiet dinner by 7:00, while the car runs to pick up the kids from soccer practice.

If vehicles can get from place to place on their own, they can spend more time doing what they're made for and less time sitting in the driveway. But if millions of families that need three cars today will soon only need one, it will be a big hit to the auto industry's bottom line.

That might only be the beginning of the disruption, however, because if cars can really move autonomously from point to point, there's no reason for them to have any downtime at all. Maybe you will participate in a network of self-driving cars, like Zipcars that drive themselves, and simply pay as you go. When you need to make a trip to the grocery store, you will summon a car using your smartphone, and in minutes it will roll up outside, ready to take you. You will leave for work every morning at 8:00, and a car will always be there waiting for you. And because there are no drivers to pay and the entire fleet can stay busy constantly, the cost of such a service could be extraordinarily low.

In fact, participating in a shared self-driving car fleet would likely be a better choice for almost everyone than financing a $25,000 car. Lawrence Burns at Columbia University's Earth Institute studied the potential for shared self-driving car systems and concluded that in a small city like Ann Arbor, Michigan, such a service would be 90 percent cheaper than car ownership, wait times for a vehicle would average about twelve seconds (two minutes at peak times), and only 15 percent of the city's current number of cars would be needed. In a place like Manhattan, Burns concluded that rides in a self-driving car fleet would cost about a dollar per trip, compared with $7.80 per trip in yellow cabs.

The service could use the smallest vehicle necessary for a particular journey: if you're riding uptown by yourself, it's much more efficient to take a two-seater.[12] Who would have thought the personalization of mass transit could make it cheaper, too?

For longer trips, bus service might become more personalized and flexible. If twenty people in your neighborhood were going to New York today, along with five people in the next town over, the self-driving bus could pick you up at your house.

With new apps, you could send your own car to pick up dry cleaning or get take-out food, to refuel or recharge in its free time, to find the nearest open parking spot and wait there, or to pick you up for appointments on your calendar. A commenter on my Facebook page imagines parental controls that will let her children use the car for approved destinations. Pizza delivery, school buses, police cars, ambulances, and trash collectors could all benefit from customized artificial intelligence in self-driving vehicles.

The widespread availability of cars dramatically changed the American landscape following World War II. Entire towns grew up dependent on the automobile but offering affordable family homes with yards and one-car garages. Suburbs, drive-in movies, McDonald's, A&W Root Beer, and Disneyland owed their success, or even their existence, to the automobile. The car became a symbol of freedom—a liberating machine for tens of millions of Americans. People could live in the country and work in the city. They could haul around their kids and their kids' friends. They could take weekend jaunts to the beach. They could pull trailers and take their families to the Rocky Mountains. The car made life richer. As Thrun told me, "I believe the car as an invention had a bigger impact on the twentieth-century society than any other invention—I think more so than television—cities look different, work looks different, social relationships look different, almost everything has been impacted by the car."

Self-driving cars have the potential to accomplish a transformation of the same magnitude. They will create unimaginable new opportunities for work and leisure, further shrink our sense of distance, change our landscape, and free up the useless hours we currently spend driving a car or sitting in traffic.

Imagine that you could roll out of bed at 5:30 in the morning and climb into the self-driving car that had just pulled up to your home. You could drift back to sleep as the car whisked you to work in the city a hundred and fifty miles away (less than an hour-and-a-half trip at 120 mph), waking just as you arrived around 7:00. Imagine that you could work throughout your evening commute, opening up time at home to spend with your family.

This scenario sounds like something from the life of a chauffeured celebrity or a Fortune 500 CEO. Some of it sounds impossible even for the superrich. But millions of average Americans could soon be living in such a world, finally liberated from the excruciating commute they once made every day.

* * * * *

All of this could happen. The technology Sebastian Thrun and others are pioneering could allow us to break out in transportation. Google engineers today are commuting to work in self-driving cars. Theoretically we could all start using them by 2020. But that almost certainly won't happen. The technological challenge of designing Priuses that drive themselves pales in comparison with the legal slog it will take to clear the way for their use. We can work our way through the thousands of regulations that stand in the way of the breakout in transportation, but if we can't overcome the prison guards who will emerge to delay the disruptive technology, it could be decades before we fully enjoy the breakthroughs that are here today.

The first and most obvious barrier is liability law. Who will be responsible if a self-driving car gets into an accident? The owner? The passengers? The manufacturer? A software vendor? What if the owner knew of a faulty sensor but did nothing? There are hundreds of hard cases that today's liability system is poorly equipped to handle.

The only states that have legalized self-driving cars formally—Nevada, California, and Florida—still require a human operator to remain at the wheel, attentive and ready to take over at an instant's notice. The laws don't even allow the person in the driver's seat to read while the car does the work. Unmanned cars are out of the question right now. Depending on the state, these requirements reduce the technology to a fancy form of cruise control.

Even if we could work out the complicated liability problems, hundreds of other federal, state, and local laws could stifle a real breakout. In many states, for instance, dense convoys of self-driving cars would violate tailgating laws. In addition to American laws, there are international laws to deal with, like the Geneva Convention on Road Traffic. All of these regulations alone could delay the availability of fully self-driving cars for decades.

If the challenge were simply the bureaucratic headache of sifting through outdated laws and rewriting them, at least we'd know what we had to do. But the legacy automakers and dealers, the Teamsters, the transit unions, the taxi drivers, the rental car companies, local law enforcement (which profits from traffic tickets), and others will not give up without a fight. Many of these prison guards, powerful interests whom self-driving technology threatens, will undoubtedly try to use the thicket of existing laws and regulations—and to erect new ones—to keep us trapped in the past. A struggle taking place today in the taxi industry is a preview of the fights to come.

Uber and Out

Ridah Ben-Amara was sitting in his Lincoln Town Car in downtown Washington, D.C., just before 9:00 on the crisp morning of January 12, 2012, when the iPhone on his dashboard buzzed to get his attention. Ridah was a livery driver filling his spare hours by picking up jobs for the fast-growing company Uber, which allows customers to summon black-car service using their smartphones and charge the trip to their accounts, no cash exchanged and no tip required. The iPhone's GPS dropped a marker on the map where his customer—labeled "Ron"—was waiting. As Ridah approached the pick-up location a few minutes later, his phone called Ron to tell him his Uber ride was arriving.

Ron got into the backseat and directed Ridah to the Mayflower Hotel on Connecticut Avenue, a popular breakfast spot for the city's elite. Ridah made his way through the capital's infamous rush-hour traffic and pulled up in front of the Mayflower. Ron got out of the car quickly and was gone. The fare would be charged automatically to his credit card, so there was no money to exchange. It was a typically seamless Uber transaction.

Until officials from the District of Columbia Taxicab Commission surrounded Ridah's car, demanding to see his taxi license and proof of insurance. "Ron," it turned out, was Ron Linton, the D.C. taxi commissioner. Washington's taxi companies weren't happy about Uber's new service, which was quickly gaining popularity and luring customers away from their dingier, less reliable, and cash-only cabs. Linton, after publicly declaring Uber to be illegal, took it upon himself to launch a "sting" operation against the company—which was operating openly and, it said, within the law.[13]

As the *Washington Post* reported, Linton's DCTC goons proceeded to slap Ridah with $1,650 in fines for four supposed violations: "Not holding a chauffeur license, driving an unlicensed [taxi] vehicle, not having proof of insurance and charging an improper fare." The *Post* quoted Linton's accusation that Uber's "improper fare" violated "a city law that says limousine trips must have a fare set in advance; Uber's system uses time-and-distance metering.... Linton said Ben-Amara refused to cite him a fare before the trip began. 'What they're trying to do is be both a taxi and a limousine,' Linton said. 'Under the way the law is written, it just can't be done.'"[14] *It just can't be done.* That's an epigraph that should be carved over the entrance of every bureaucratic prison of the past in the country.

Linton followed classic prison guard tactics in using his interview with the city's paper of record to warn potential Uber drivers that they could be next: "It kind of serves as a message to the others that are doing this, they're not going to be immune," he warned.

Ridah Ben-Amara was an early casualty in what has become an ongoing war between Uber and the D.C. Taxicab Commission, which enjoyed the support of the city council. Over the next year and a half, this alliance of prison guards sought to impose on Uber a minimum fare that was five times that of a taxi, to disqualify vehicles less than 3,200 pounds (targeting Uber's hybrid cabs), to require credit-card processing through the meter (which Uber's cab partners could not do), to require the company to share data about its rides with the DCTC every twenty-four hours, and to ban any operator with a fleet smaller than twenty vehicles (many Uber drivers work independently).[15] These proposals were intended to undermine Uber's operation in the District. Most were defeated after aggressive grassroots campaigns organized by Uber. (The company has been remarkably successful at harnessing its loyal customer base to defend itself politically.)

Uber has faced the same fight in nearly every city it has entered. Prison guards in city governments and taxi commissions consistently abuse their power to protect the taxi companies from competition. In New York, regulators forced Uber to shut down taxi service last year, and lawsuits from the city's livery companies have complicated its efforts to return.[16] In Las Vegas, Uber ran up against a regulation that established a forty-dollar minimum rate for all livery services.[17] In Chicago, the taxi companies sued Uber for trying to "cash in on [their] good name" and for allegedly confusing customers (though Uber said it had received no complaints).[18] The Boston suburb of Cambridge said the company could not use GPS to meter rides until the National Institute of Standards and Technology set technical guidelines for the ubiquitous location service (which is accurate to within a couple of feet).[19] Uber's vocal customers defeated most of these efforts as well.

Uber has been the most prominent challenger of the taxi monopoly, but the prison guards have been equally vicious in trying to stamp out similar startups like Hailo and SideCar. FlightCar, a service that allows travelers at the San Francisco airport to rent their cars to incoming visitors while they're away, was hit with a lawsuit in June for failing to comply with all the regulations that apply to the big rental car chains like Hertz and Avis.[20]

Tesla, the luxury electric car company led by PayPal cofounder Elon Musk, is under regulatory assault in a number of states for trying to sell its cars directly to consumers over the internet rather than through dealer franchises as the major car companies do. The *Wall Street Journal* reported: "The focus of the power struggle between Mr. Musk and auto dealers is a thicket of state franchise laws, many of which go back to the auto industry's earliest days.... Dealers say laws passed over the years to prevent car makers from selling directly to consumers are justified because without them auto makers would use their economic clout to sell vehicles for less than their independent franchises."[21]

In other words, the dealers want to protect the regulatory privileges that allow them to charge customers (us) more than is necessary. Tesla's attempt to sell its cars directly over the internet (or through company-owned stores) broke the prison guards' rules, so they're trying to kill it.

You might have noticed that Uber actually operates a lot like the shared fleet of self-driving cars I described earlier—customers summoning the vehicles with their smartphones and being picked up a moment later. In fact, take away Uber's drivers, add some sensors, and stuff a few computers in the trunk and you've pretty much got the model. And in Ron Linton and his taxi commission, the prison guards have their model too.

Stalling Out on Innovation

Once industry pioneers perfect the software for self-driving cars, which Sebastian Thrun estimates could happen within a couple of years, they will also have to get past the most imposing prison guard in the automotive industry: the National Highway Transportation Safety Administration (NHTSA, pronounced "*nit*-suh"), which regulates cars, trucks, motorcycles, and anything else on the road. Every vehicle must obtain the NHTSA seal of approval before it can be sold to the public.

"Most people have no idea of all the aspects of the car that are controlled by federal regulation one way or another, and innovations that are stymied on cars for one reason or another," says Robert Norton, a former assistant general counsel for Chrysler. Norton spent much of his career negotiating with NHTSA on behalf of the auto companies, including stints affiliated with each of the American Big Three. Most recently he was chief class action and vehicle regulatory counsel at Chrysler, a position that for seven years gave him a close-up view of how federal regulators kill innovation.

Consider the relatively small detail of a car's headlights. "Headlights on cars were originally mandated to be round," Norton told me. "When the industry wanted to make them square for a while, when that was in vogue, they had to go in and say, 'Mother may I?' and give NHTSA very specific details about exactly how the headlights would work." Eventually permission was granted, but the agency continues to block advanced lighting technologies that would improve safety. The relevant bureaucrat inside NHTSA "has for a long time been affectionately referred to as 'The Lighting Nazi,'" says Norton, for his fixed views about "the archaic laws and how it should be going forward."

"Now—today as we're sitting here—there are technologies available on cars in Europe that make lights safer. But the United States government does not recognize that technology yet, so the cars that are sold in the U.S. have to have the headlights that are made in Europe removed and have basically dumbed-down headlights put in for the American market." Norton cites the example of Audi, which in Europe uses cameras automatically to dim individual LEDs in the headlight to avoid blinding other drivers while still illuminating the rest of the road and uses GPS to bend light around curves before the driver even turns the wheel. Volvos have a shutter that can automatically adjust to keep lights from blinding other drivers and open back up to full brightness when no cars are in view. "Our government is just not moving yet," he says. "They're holding back that innovation though it's already out on the market."

At one point in Norton's career, a design team wanted to use clear LEDs on the rear of the vehicle that would change color as needed (for instance, red when you stepped on the brake pedal). The manufacturer argued that LEDs, which never burn out, would be safer for consumers. "At that time the government would really hear nothing of it," Norton says. "NHTSA was absolutely refusing to let us have them."

NHTSA enforces its will in large part through crash tests, which at first glance seem sensible. But those tests, complains Norton, are arbitrary and to a large extent meaningless. The front-impact 35-mph crash test presumes drivers are unbelted, even though seat-belt usage is approaching 90 percent in many states. It is a forty-year-old notion of how people ride in automobiles that has never been revised. Even the Insurance Institute for Highway Safety uses a side-impact test to rate crashworthiness that "specifically is designed to imitate the front of a previously manufactured Ford F150 pickup truck," which at that time represented the most popular vehicle of that size in the United States.

"It's a little bit comical," Norton says, "when you really know how these tests are done. If you don't happen to get hit by a Ford F150 at that exact speed and angle, well, then all bets are off. Any scientist will tell you we have no predictor of how accurate this test is for, say, a Dodge truck, hitting you at 40 mph. What it really means for real-world safety, no one really comments too much about."

In fact, the latest series of crash tests require crash dummy chest-deflection measurements that are more accurate than the calibration requirements of the dummies can even account for. For example, manufacturers are required to measure for only 15-20mm deflection in certain crashes, but the calibration of the dummy only gets as close as 75mm deflection. So the same test run with different dummies yields different results.

Like other federal bureaucracies, NHTSA's power has expanded over the years, and now it pursues a muddled, often conflicting set of priorities that have dramatically changed American cars, not always for the better. The agency "started off all about safety and then it drifted into fuel economy in the '70s with the oil embargo, and then into emissions and greenhouse gases," Norton observes. The irony is that the regulatory agency that was supposed to ensure passenger safety now demands that "we put people in very light-weight cars" to reduce emissions and increase fuel economy, "which means cars that we know are less safe. It's funny how we've come full circle."

This mission creep has left the prison guards with the power to control almost every aspect of Americans' automobiles. NHTSA now sets so many rules and tests that there is only one way to meet them. All new cars on the road "pretty much start fitting in the same box," Norton says. "You hear most of the lay people say, 'Gee, they all look so much alike these days.' Well, that's why. They're matching the government recipe." The regulators have slowed automotive innovation to a crawl.

Asked if there was nothing the carmakers could do to get around the rules, Norton laughs, recalling Chrysler's experience with its PT Cruiser, a smaller, fuel-efficient car with a retro design. It took a lot of finesse to get that retro design through the government regulations, he says, including one feature that the company wanted to include: tinted glass. "What Chrysler had to do was actually have the vehicle meet the government regulations that were intended for minivans, multi-purpose vehicles, in order to be able to put tinted glass in the side windows." Minivans, apparently, are allowed to have tinted glass but passenger cars are not—a regulation that dates back to the 1960s. "So Chrysler beat the scheme and put out tinted windows in the PT Cruiser, and NHTSA was irritated about it," he said. "They didn't take it lying down. They caused us a lot of grief and aggravation."

Auto companies don't often dare to play games with the prison-guard regulators, however. When they can throw the rulebook at your core business, you don't pick fights. "There's so much soft power that NHTSA has over the industry," Norton explains, "because you're always needing extensions and exemptions and 'Can I have sixty more days to give you this report?' And generally you are expecting to get the 'Mother-may-I' permission on that." But not always. At any time, he says, "if they get really irritated at you, they say, 'No, actually you can't. We want this now, and we're not going to look at this, and we're not going to consider that.' So you really are encouraged to play ball because you're counting on them for your existence." Norton describes one occasion on which an automaker spent more than three years and several million dollars working with NHTSA on an issue it knew was never a problem before reaching a "compromise" to let the agency save face.

What about self-driving cars? How will the agency that blocks changes to headlights and tinted windows respond to computers that can drive two tons of machinery down the highway perfectly? "There's a notion within the agency that because there is cyberterrorism that could take over the controls of your car and drive it into the concrete buffers, we probably shouldn't have any more of these computers taking over for people," Norton says. "I understand it's not the temperament of the current regulators to want to go very far or very fast down that road."

After all, if the computers could do it perfectly, there wouldn't be much need for NHTSA, the prison guards of obsolete auto safety requirements.

CHAPTER SEVEN

BREAKOUT IN SPACE

In 1976, four years after Apollo astronauts had departed the moon for the last time, a Princeton University physicist named Gerard O'Neill published a manifesto called *The High Frontier*, a detailed vision of humanity's near-term future in space. He imagined cylindrical space stations miles in diameter, with artificial gravity and landscapes like Southern California, complete with trees, lakes, and flowing streams. Thousands of people would live in traditional homes clustered in villages and towns. There'd be restaurants, movie theaters, playgrounds, and neighborhood barbeques. When space-dwellers sat "outside" on their decks and looked up at night, they'd see not stars but the sparkling lights of distant towns across the cylinder, hanging upside down from their perspective.

O'Neill was convinced that his designs could be commercially viable. His "islands" weren't just vacation spots; they were primarily to be manufacturing hubs, with plenty of jobs for extraterrestrial expatriates.

Residents could do manufacturing work that would be impossible on Earth, such as building extremely light structures or novel pharmaceuticals that require zero gravity to form. O'Neill also envisioned massive steam turbines designed to remain in space, powered by the sun twenty-four hours a day. The first space stations, he speculated, would be built by energy companies to generate low-cost electricity, which would be beamed back to Earth in the form of targeted microwaves. Other people on the islands would work in mass agricultural production, which would have extremely high yields at always-sunny points in space. A small local economy would grow around these primary industries, he predicted.

But how to build these gigantic cities in space? Men could mine the raw material at a base on the moon, O'Neill envisioned, after which they'd use what amounted to a huge magnetic cannon to launch it through the almost-nonexistent lunar atmosphere to construction sites somewhere between the earth and the moon.

O'Neill's ideas were more than *Star Trek*–style science fiction. He, along with a substantial number of colleagues, produced detailed engineering plans to back up the concept. He demonstrated the feasibility of a number of the ideas in the real world, such as the "mass driver" he wanted to build on the moon. There was no theoretical reason why the stunning islands he described should not be possible, even profitable. And he thought it would all happen by 2010.

The High Frontier was an instant hit, exciting Americans about space for the first time since the Apollo missions. *60 Minutes* covered O'Neill's proposals, as did the *New York Times*. As Greg Klerkx writes in *Lost in Space*, "After the book's publication, O'Neill became a media darling, appearing on talk shows, writing for newspapers and magazines and speaking before congressional hearings on the merits of space manufacturing, space solar power and space settlement. His allies included... California's governor and presidential candidate Jerry Brown, and the powerful congressman Morris Udall, who was sufficiently impressed by the energy ramifications of O'Neill's plans to lobby (albeit unsuccessfully) the Federal Energy Research and Development Administration to fund further design studies of O'Neillian concepts.... Even Carl Sagan became an O'Neill supporter."[1]

To many Americans (including me), the O'Neill vision for America's next act in space seemed a real possibility. After all, men had just walked on the moon in the same century as they had invented the airplane. And perhaps it *was* possible—as recently as the year 2000, the famed physicist Freeman Dyson predicted the concept could be viable by 2050.

Fast-forward thirty-seven years from O'Neill's proposal, to 2013. The space shuttle, which was still on the drawing board when he published *The High Frontier*, is now literally a museum piece. The United States has no spacecraft capable of sending human beings into orbit. Since the shuttle retired more than two years ago, NASA has been forced to pay the Russians for seats on their Soyuz capsules (the earliest of which launched in 1967). The International Space Station is a timid and increasingly confused project with no clear mission, and NASA itself seems aimless, lacking a vision for the future.

Bureaucratizing Space

When in my presidential campaign I advocated a manned base on the moon—a goal I have supported for my entire career—many in the media and in my own party howled with laughter. Yet building a moon base had been official government policy through most of the Bush administration and for the first two years of Obama's presidency, until he canceled the project in 2010 following ludicrous cost overruns in the early stages.

What happened? How did Americans go from being thrilled by the idea of thousands of people living and working in space, a country that celebrated visions of O'Neill's space islands on *60 Minutes*, to scoffing at the idea of sending men back to the moon? The Chinese weren't laughing. They were busily pursuing a moon shot of their own. What could be to blame for such a narrowing of imagination and increase in timidity in America?

The answer, quite simply, is NASA. The agency was once almost synonymous with the future, but in the four decades since the moon landings, it has become one of the government's most tragic prison guards of the past.

In the past eleven years—since the *Columbia* disaster led President Bush to retire the space shuttle—we have spent roughly $150 billion on NASA and the civilian space program. We have spent additional money on defense aspects of a space program. Yet today the United States, on its own, cannot launch a single human being into space. NASA has accomplished some difficult feats in its history, but spending $150 billion on the space program without developing a rocket and spacecraft to launch astronauts into orbit is near the top of the list.

NASA has reached this point by achieving a perverse breakthrough: the bureaucratization of space. The modern NASA is so risk-averse and so heavily burdened with safety processes, management, political meddling, and institutional inertia that it takes decades for new programs to get off the ground.

Only seven years and five months passed between John Glenn's Mercury 6 mission in February 1962 (when he became the first American to orbit Earth) and Neil Armstrong and Buzz Aldrin's moon landing in July 1969. In that time we figured out how to perform frequent launches, keep men alive in space for weeks, conduct spacewalks, rendezvous and dock two spacecraft in orbit, travel to the moon, land on it, walk around the lunar surface, launch from the moon, and return to Earth. Each of these achievements presented innumerable challenges, all of which were overcome in well under a decade.

The shuttle program lasted thirty years, not counting the decade in which it was developed. And after those thirty years, we are reduced to buying seats for American astronauts on a class of Russian spacecraft first launched almost half a century ago. What further evidence do you need that bad bureaucracy can set breakthroughs back generations?

NASA has conducted some spectacular science and robotic missions in recent years, including the Mars rovers *Spirit*, *Opportunity*, and *Curiosity*, and the agency deserves credit for these achievements. But even if rocket scientists and astrophysicists think in longer time frames than most people, it would be desirable for the human space program to make some important advances over the span of their entire careers.

The golden age of the space program makes all Americans feel proud. But today's non-manned interim diminishes the luster of that history and disappoints even NASA's employees and supporters. It's obvious that the bureaucratic model is failing, and failing expensively.

Freeman Dyson writes of the two NASAs: the "real NASA," which he describes as "intensely conservative" and "dedicated to preserving existing programs," and the "paper NASA."[2] Klerkx elaborated on this observation: "The 'paper NASA' has long been engaged in maintaining an image that is bold, daring and committed to a *Star Trek* kind of future," he says, "... detailing fantastical futures that are always, so it is promised, right around the corner.... Impressive as such efforts may seem, nothing in the 'paper NASA' process costs very much and few of the recommendations that emerge from these blue-sky activities ever see the light of day."

"Such remarkable and visionary things require money," Klerkx continues, "which is controlled and administered solely at the discretion of the 'real NASA,' which is all about the convoluted bureaucratic mix of political back-scratching, government appropriations, industrial contracts and—at the core of it all—the one inarguably tangible result of their interplay: jobs. The 'real NASA' is concerned with keeping the dollars flowing from its immediate and long-tested sources. Where they flow to within the agency, as long as they do, is of considerably less importance."[3]

The "real NASA" is extraordinarily risk-averse in order to protect this stream of funding, the opposite of what we want from an agency that's supposed to be at the forefront of discovery. This is a familiar disease in any large bureaucracy, but for some reason it's a little harder for Americans to believe that NASA, the agency behind moon landings and the Hubble Space Telescope, is just another bureaucracy. We don't want to believe that they often act more like IRS agents than intrepid explorers.

This bureaucratic mess is the main cause of the agency's history of horrendous cost overruns. The shuttle, for instance, was originally supposed to fly dozens of times per year at a cost of about $50 million per flight. Ultimately the program delivered far fewer flights than originally planned at a cost of roughly $1.5 billion each.[4] That's almost thirty times the early estimates. But when everybody has to have a piece of the program, costs pile up.

In addition to the absurd operational complexities that NASA creates for political reasons, the agency pays its contractors (the giant aerospace companies) on a "cost-plus" basis, meaning *whatever they spend* plus a markup. This gives the companies a disincentive to save money. Seen in this light, the overruns are much less perplexing.

Rather than trying to fix this perverse incentive by changing its payment structure and introducing greater competition, NASA has protected itself and the coterie of contractors it works with from any challenge to their privileges. *Lost in Space* details NASA's determination to kill the purchase of Russia's aging Mir space station by a group of private entrepreneurs who would have put it to commercial use. The deal was all but done, but NASA pulled out all the stops to prevent it from happening, apparently because the private venture would have competed with the new International Space Station.[5]

While NASA obstructs innovation by private-sector space entrepreneurs, its major partners have continued to win contracts even after repeated and expensive failures. In 1998–1999, Lockheed Martin underwent five formal investigations for mismanagement. Its errors caused the loss of a spy satellite ($1.3 billion), two military satellites ($800 million and $250 million), the Mars *Pathfinder* ($265 million), the Mars climate orbiter (Lockheed's famous failure to convert from English to metric units—$145 million), and the Genesis probe ($264 million). Yet after this string of catastrophes, NASA rewarded Lockheed with a $7.5 billion contract to build the Orion capsule, its next vehicle for manned spaceflight.

Lockheed is not alone in its expensive failures. Almost all of NASA's traditional contractors are gigantic bureaucracies themselves. The agency's endless red tape and paperwork mean that only a few large companies have the resources even to apply. Few entrepreneurs or private space startups could cope with the costs, complexities, or timelines involved in winning NASA contracts, and they are no match for the traditional aerospace companies' massive lobbying operations. This is not an accident.

Congress deserves a share of the blame for treating NASA as a political football, which has made the space program much more expensive than it would otherwise have been. Many of the agency's strongest supporters in Congress have NASA centers in their districts or states, and the centers themselves are astute lobbyists for a piece of the action. Many of NASA's activities, therefore, are driven by politics, not by the needs or interests of the space program.

In 1984, I predicted in my first book, *Window of Opportunity*, that "the longer we keep space a government monopoly, the greater the danger of smothering initiative."[6] Thirty years later, NASA is still a government monopoly, and that's exactly what has happened.

The Prize Model

There is a way forward for America's space program. Despite the prison guards' efforts to preserve the status quo, we have begun to see exciting progress outside of NASA. Genuine pioneers are opening space to the private sector, taking risks—both financial and physical—in pursuit of the high frontier. They've done it with encouragement not from NASA but from *prizes*.

Prizes have figured prominently in the history of aviation from its infancy. Starting in 1906, Britain's *Daily Mail* offered rewards for the first aviators to achieve various milestones, including a nonstop flight between London and Manchester and flying across the English Channel. In the United States, William Randolph Hearst offered $50,000 in 1910 to the first person to fly from coast to coast in thirty days. Most famously, Raymond Orteig offered a prize of $25,000 in 1919 for the first person to fly nonstop from New York to Paris. It took eight years, but Charles Lindbergh won the Orteig Prize in 1927.

These competitions were far more dangerous than many today might imagine. In the 1927 Dole Air Race from California to Hawaii, only two of more than fifteen entrants made it to their destination. But the pilots took such risks eagerly and freely, and in doing so made enormous strides in advancing and popularizing aviation.

A prize system similar to that of the early twentieth century, this time aimed at enticing private companies to pursue our goals in space, has produced far more progress in human spaceflight over the last ten years than NASA has done over the same period. It also better reflected our values of entrepreneurship and adventure than did the massive bureaucracy incompetently managed by Congress and appointed bureaucrats.

In 1996, the privately funded X Prize Foundation offered a comparatively small $10 million prize for two manned suborbital flights in a reusable spacecraft within two weeks. It drew more than two dozen competitors, who worked independently on a human-rated spacecraft. Just eight years later, in 2004, a group led by Burt Rutan and Microsoft cofounder Paul Allen won the Ansari X Prize with SpaceShipOne, an inventively designed craft launched from a jet plane at high altitude.

SpaceShipOne was a new beginning for the commercial space industry. The pioneers behind the vehicle joined with Richard Branson in the Virgin Galactic partnership to produce the larger SpaceShipTwo. It will carry eight people on half-hour trips into suborbital space beginning in 2014. The initial cost of a ticket will be $200,000—a lot of money, to be sure, but a far cry from the $20 million that Dennis Tito paid for a trip to the International Space Station in 2001.[7] There might be thousands of people willing to pay Virgin's price. Indeed, there's already a waiting list.

SpaceShipOne, the first private manned space vehicle, and Virgin Galactic, the first private manned spaceflight carrier, are almost unimaginable in the absence of the Ansari X Prize. The contest's successor is the Google Lunar X Prize, which will award $20 million, and up to $10 million in bonus prizes, to the first group successfully to land a robot on the moon, beam high definition video back to Earth, and travel five hundred meters. Twenty-three private initiatives around the world, including seven led by American teams, are now working to land robotic spacecraft on the moon once again—with minimal government funding.

I met with the leaders of one Lunar X Prize team, Penn State's Lunar Lion, who told me their current estimated cost to launch is only about 10 percent of their original expectation and a tiny fraction of what NASA would pay for a similar project. We hosted a discussion of the Lunar Lion project, which you can see at www.BreakoutUniversity.com.

These prizes have jump-started the commercial space industry, which now includes dozens of companies with promising business models.

SpaceX, another company founded by Elon Musk, created the Falcon 1 and Falcon 9 rockets as well as the Dragon spacecraft and now runs cargo resupply missions to the International Space Station. Within a few years, its vehicles may carry astronauts as well, and its planned Falcon Heavy should be capable of providing enough lift to reach Mars. NASA deserves some credit for taking such steps toward reliance on commercial space services in recent years, and so does the Obama administration, which pushed the agency further in this direction.

Other exciting developments include Space Adventures, which is offering to launch two people on a lunar flyby trip. Another company, Golden Spike, is working to carry human passengers to the moon for lunar landings by 2020—again, for a small fraction of what NASA would pay for the same trip. (I am an unpaid member of Golden Spike's board of advisors.)

This boom in commercial spaceflight portends an exciting future for America in space. We are finally beginning to realize Gerard O'Neill's dream of a commercially practicable space industry.

★ ★ ★ ★ ★

The success of competitive prizes like the Ansari X Prize points us toward the best way for government to support big breakthroughs in a whole range of areas, at a far lower cost than our current programs.

Instead of spending almost $20 billion each year on the space program and getting disappointing results, what if Congress set aside a large sum for prizes—say 10 percent of NASA's budget, or $18 billion over a decade? We could save hundreds of billions and get better results. We could reduce the size of NASA and refocus its mission on breakthroughs in science and technology rather than developing or operating basic launch vehicles and spacecraft.

After I discussed the prize concept (which I have advocated for most of my career) with Robert Zubrin, an expert on space exploration, in the 1990s, he estimated in his book *The Case for Mars* that if Congress posted "a $20 billion reward to be given to the first private organization to successfully land a crew on Mars and return them to Earth, as well as several prizes of a few billion dollars each for various milestone technical accomplishments along the way," it would draw numerous competitors. The actual mission, he thinks, could cost as little as $4 billion, leaving the winner with a $16 billion profit and the taxpayers with a system that gets to Mars for a fraction of NASA's annual budget.[8]

Prizes have several huge advantages, which Zubrin also points out. First, taxpayers don't pay anything until we get results, and we never pay more than the prize amount. If no one offers a system of launch vehicles and spacecraft that meet the prize specifications, *it doesn't cost anything*. And cost overruns are impossible even if there is a winner. After spending $150 billion on NASA for no current manned capability, this is quite a virtue.

Second, prizes would result in systems that are radically cheaper than those NASA has produced. Unlike the agency's "cost-plus" contract basis, in a prize system, a company has to raise or borrow every dollar it spends, which means they use their money efficiently. Many competitors will spend money investing in technology and developing new solutions but won't win the prize. And they spend all the money before the taxpayers ever have to pay anything.

Finally, Zubrin points out that competition breeds better results. While NASA projects typically produce only one working design, a single prize incentive could result in several designs that make it to the flight stage, each with different merits. Awarding runner-up prizes further stokes the competition.[9] With such a prize-based, entrepreneurial approach, we can recapture the spirit of adventure and again be the envy of the world in space.

We could use prizes to achieve critical breakthroughs in other important areas of life. In 2012 the federal government made $108 billion worth of improper payments—money it simply never should have spent. If we had added that amount—just one year's waste—to a competitive prize fund for big breakthroughs, we might find cures for Alzheimer's disease, diabetes, or cancer. We might get an online learning platform that teaches students grades K–12 twice as well in half the time. Or a reproducible set of lab-grown lungs. Or a car that drives from New York to Los Angeles with no human being at the wheel. The list is endless. And again, *none of this money would ever be spent unless someone actually achieved a breakthrough.*

What kind of world might we already be enjoying if we had set aside even a fifth of the $831 billion stimulus for breakthrough prizes? At worst, we would be $166 billion richer right now. At best, we would be talking about the new American breakout rather than the new normal.

CHAPTER EIGHT

BREAKDOWN IN GOVERNMENT

Only 248 people lived in Boynton, Oklahoma, when state officials showed up in 2010 with federal stimulus dollars burning a hole in their pocket. Washington had sent $16 million to the state to improve its sidewalks as part of the "economic recovery" package, and evidently the Oklahoma Department of Transportation had run out of better places to build them. Perhaps they thought a gleaming new stretch of sidewalk "infrastructure" in rural Boynton would set off an explosion of economic activity. Perhaps they thought Boynton was so remote that they could hide waste there without anyone's noticing—a sort of Yucca Mountain for the fiscally toxic byproducts of the stimulus legislation. Either way, residents watched bemused as contractors arrived to tear up the existing walkways and replace them with new ones, including a sidewalk that led into a ditch.[1]

"This was brand new sidewalk, five years old," resident Mike Lance told a local TV station about one stretch of walkway. "I mean, it didn't even have weather stains on it.... This sidewalk looked just like this before they started."

"It's goin' to nowhere," he exclaimed, pointing to another section "that fronts no homes or businesses, and leads directly into a ditch."

"If I'm wrong, convince me," he said. "If I'm wrong, I'll say I'm sorry...but I just don't see me saying I'm wrong on this project."[2]

Boynton wasn't the only town thus "enhanced" with stimulus money. Council Hill, Oklahoma (population 129), received $245,000 for "pedestrian improvement," about $1,900 for every person living there.[3] Other towns across the state got similar infrastructure upgrades.

The sidewalks to nowhere in Oklahoma (as well as others in Michigan and Florida) are just one example among hundreds that Senator Tom Coburn has cataloged in thirty-five incredible reports on government waste. His annual "wastebook"—a must-read in Washington—would be funny if hardworking taxpayers weren't financing every boondoggle he finds.

Senator Coburn points to at least $1 billion in government payments to more than 250,000 dead people in the last decade,[4] $114 billion worth of tax breaks to millionaires (including for their gambling losses, rental expenses, and electric vehicles),[5] and $2.5 billion in improper food stamp payments.[6]

Those are just a few of the larger-ticket items. Senator Coburn's books contain hundreds of smaller but equally absurd expenditures, such as $27 million for pottery classes in Morocco (a country that has been making pottery for thousands of years),[7] $6.5 million in overpayments for brand-name drugs with generic alternatives,[8] $3 million for a children's golf program in South Carolina,[9] $5 million for a pedestrian bridge between the Saint Louis Zoo and its parking lot,[10] and $2 million a year in service fees on unused federal bank accounts that have a zero balance.[11]

Taxpayers spent $325,000 for a robotic squirrel to test how snakes responded to it,[12] $300,000 to promote the consumption of caviar,[13] $1.5 million for a NASA video game mimicking a trip to Mars,[14] $500,000 on a pet shampoo company,[15] $765,000 to build an IHOP in Washington, D.C.,[16] $484,000 to build a Mellow Mushroom pizza restaurant in Texas,[17] $937,000 to produce an online soap opera starring Billy Dee

Williams,[18] $100,000 to analyze *World of Warcraft* gamers,[19] $1 million to display snippets of poetry in zoos,[20] $90,000 on a Shrek-themed campaign promoting Vidalia onions,[21] $100,000 promoting Idaho's wine industry,[22] $30,000 for a Hawaiian cookbook,[23] $700,000 to study greenhouse gas emissions from cow burps,[24] and $10,000 on talking urinal cakes to fight drunk driving.[25]

Senator Coburn's catalog goes on, and on, and on. Each example is more absurd and extravagant than the last. He has compiled thousands of pages of incontrovertible proof that government is broken—badly.

Some of this is simply incompetence, and some of it is willful. But even what begins as incompetence ends up staying in place intentionally, as the prison guards refuse to consider defunding wasteful programs.

The waste is more than an accumulation of small-ticket items, each one amounting to tens or hundreds of millions of dollars. The bureaucracy is often wasteful on a grand scale. The Government Accountability Office identified nearly $108 billion in improper payments in 2012—more than a trillion taxpayer dollars in ten years that never should have been spent.[26] Yet under the current system, there is no hope of a remedy.

The massive federal bureaucracy that presides over this waste is utterly incapable of containing it, and yet the bureaucrats spend their time creating and enforcing tens of thousands of regulations that touch every aspect of our lives. Congress has passed increasingly vague laws in recent decades, ceding the responsibility to write rules to the bureaucracy and giving these unelected federal employees more power.

Of course, a certain amount of regulation is necessary. Somebody has to apportion the wireless spectrum, reserving frequencies for emergency services and making sure your home Wi-Fi doesn't interfere with your neighbors' broadcast TV. The government properly sets the rules of the road for motorists. The regulatory state we have today, however, surpasses these public purposes by many orders of magnitude. The vast majority of regulations on the books are just as absurd as the wasteful spending programs in Senator Coburn's reports. Indeed, the costs of unnecessary regulation far exceed those of the worst-managed government programs.

The *Code of Federal Regulations* exploded from 19,000 pages in 1949 to almost 170,000 pages in 2011, an eightfold increase.[27] Do Americans really need eight times more controlling today than when I was a child? In the past decade alone, the *CFR* grew by more than 20 percent.[28] Between 1993 and 2012, the government added eighty-one thousand pages of new rules to the *Federal Register*.[29] The regulations that came from the Dodd-Frank Act would fill twenty-eight copies of *War and Peace*, and rules issued by bureaucrats took "13,789 pages and over 15 million words,...which is equal to 42 words of regulations for every single word of the already-hefty law, spanning 848 pages itself," according to one analysis.[30] It's completely out of control.

The ridiculous rules are enormously costly. The Competitive Enterprise Institute estimates the cost of compliance at more than $1.8 trillion a year, or over half the dollar value of the federal budget.[31] That number exceeds the entire GDP of Canada or Mexico.[32] For a family making $49,705 a year, the annual cost of regulations embedded in their daily life is $14,768, or 30 percent of their household budget.[33] Businesses, too, pay a high price. The Small Business Administration estimated in 2010 that a company with twenty employees or fewer paid $10,585 per person in regulatory compliance alone.[34] In 2012 the federal government added 854 new rules affecting small businesses.[35] All of that regulation is a hidden tax on American families.

Even worse, the stacks of unnecessary rules strangle innovation. Over and over in this book, we've seen examples of pioneers struggling to overcome the law just to have the chance to compete. Incredible innovations—Khan Academy and Udacity, Dr. Atala's regenerated kidneys, Uber's car service and Google's self-driving cars, George Mitchell's and Harold Hamm's fracking boom—are all under assault by prison guards who use the regulatory state to protect the past. And these are just the pioneers who are still succeeding. The unbearable burden nips untold thousands of innovations in the bud. Obviously, this enormous headwind of regulation—about four thousand pages of new rules every year—slows economic growth substantially.

What if all the accumulated government regulations shaved just a couple of percentage points off of economic growth each year? It's not hard to believe. Kill a few Microsofts or Amazons (or, for that matter, Fords or FedExes) while they're still operating out of garages, and it's a much poorer country.

It turns out this is exactly what overregulation has done. A study published in 2013 in the *Journal of Economic Growth* finds that the sixfold increase in total regulation from 1949 through 2005 cut growth by an average of two percentage points a year. Over time, those two percentage points add up to a lot that we're missing out on. By the end of that period, the researchers concluded, annual GDP in the United States was just "28 percent of what it would have been had regulation remained at its 1949 level."[36]

What does that mean for average Americans? As *Reason* magazine translated the findings, "Federal regulations have made you 75 percent poorer.... The average American household receives about $277,000 less annually than it would have gotten in the absence of six decades of accumulated regulations—a median household income of $330,000 instead of the $53,000 we get now."[37] If the real cost of overregulation is even half that number, we would be talking about a typical American family's earning $115,000 a year.

The economist Alex Tabarrok has another way to think about this. "In terms of innovation," he says, "if productivity had continued to grow along the 1947–1973 trend then we would be living today in the world of 2076 instead of the world of 2011."[38]

This is the real cost of overregulation: the unseen breakout that should have been.

Of course, this is not to say we should get rid of all regulation. But a median household income of $115,000 would certainly solve a lot of problems. The American people cannot allow the prison guards of the past, the forces behind most of the 80-percent increase in regulation since 1949, to continue dragging us—their prisoners—further and further behind.

Breakdowns in the States

It's not just the federal government that's breaking down. Many state and local governments, too, are falling apart in places where they have been disastrously mismanaged. Government at all levels is failing us. Meredith Whitney, a financial analyst, offers a startling example of the problem in her book *Fate of the States*. In the early 2000s, she recounts, the head of the Contra Costa County, California, firefighters' union "negotiated a sweet new contract for his members." The contract provided that "veteran firefighters could now retire at age fifty with an annual pension equivalent to 90 percent of their salary." This deal was a great boon for the firefighters but not for the taxpayers.

Predictably, the weight of the pensions proved too much for the county to sustain. By 2012, in the wake of the recession, Contra Costa County could no longer afford to keep all its firehouses open and meet its pension obligations at the same time. When the director of the local taxpayers association looked into the matter, she found that there were 665 retired county employees with an annual pension of $100,000 or more and twenty-four who earned over $200,000, even in retirement. As Whitney puts it, "Everybody might still love firefighters, but what they did not like was retired fifty-five-year-olds taking home $100K a year at a time when many taxpayers were out of work and could not afford to put any money aside for their own retirements."[39]

The newspapers these days overflow with similar examples of excess. Whitney describes a California prison guard (the literal kind) who "with a base salary of $81,683 collected $114,334 in overtime and $8,648 in bonuses." This man "was eligible for an annual $1,560 'fitness' bonus for getting a checkup," and he "could retire at fifty-five with 85 percent of his salary and medical care for life."[40]

Such reckless management and outright corruption has left states broke. Illinois has roughly $100 billion in pension obligations, the vast majority of which is unfunded, and another $55 billion in unfunded obligations for retiree healthcare.[41] In California, the debt story is even more alarming. State and local governments in the Golden State carry a $1.1 trillion debt.[42] Yet California continues to spend big on goodies like a hundred-billion-dollar high-speed rail line from San Diego to Sacramento.[43] Many of the major states—California, New York, and Illinois among them—behave just as recklessly as the federal government.

A Broken-Down Machine

Regardless of their political ideology, Americans see with increasing clarity that they cannot rely on government to fix our country's biggest problems. In fact, it would be hard to look at the current government and think that it is capable of doing anything big at all. It's failing before our eyes.

I first encountered the idea that the machine of government is simply breaking down during a research trip in 2000 for the Hart-Rudman Commission on National Security. We were visiting at the Harvard Kennedy School of Government with a former senior leader in the Pentagon. Like most of these commissions, we were focused on strategy and policy.

Suddenly this experienced official said, "You know, the real problem is that the machine doesn't work." We paid close attention as he went on: "Everyone in Washington wants to debate policy. No one wants to explore implementation. We get into huge fights over the right policy, but then nothing happens. It is as though the steering wheel is disconnected from the car. All the politicians fight over turning the steering wheel right or left, but in the end it doesn't change the direction of the car, because it isn't connected to the wheels."

In the years since that conversation, it has become even clearer that the machinery of government isn't responding. The system is coming apart.

Whatever we decide we want the government to do, we want it to be able to *do* it. Yet today our bureaucracies and our systems are so ineffective, incompetent, and obsolete that—other than on the battlefield—we can't get anything done.

This problem afflicts every level of government—from cities, counties, and school boards to state capitals and on to Washington. In some respects the problem is nonpartisan. We saw it with the federal response to Hurricane Katrina. We have seen it more recently in Detroit.

Indeed, the bankruptcy of the city of Detroit is a symbol of the steady decay of government competence in America. It's horrifying that the city that had the highest per capita income in the country in 1950 could collapse economically, culturally, and socially in just a few decades.

Detroit's population has been cut in half—from 1.8 million in 1950 to 700,000 today.[44] The collapse in population has left seventy-eight thousand empty houses. Some are available for one dollar (yes, it's true), but no one will buy them.[45]

The number of manufacturing jobs in Detroit dropped from 296,000 to 27,000 over the same period.[46] No one will create new jobs in the city because it has become a public safety and public services wasteland. In some ways it resembles the devastated world of *Mad Max*.

Bill Nojay, who served as chief operating officer of the Detroit Department of Transportation in 2012, described his experience in the *Wall Street Journal*:

> I was hired as a contractor for the position, and in my eight months on the job I got a vivid sense of the city's dysfunction. Almost every day, a problem would arise, a solution would be found—but implementing the fix would prove impossible.
>
> We began staff meetings each morning by learning which vendors had cut us off for lack of payment, including suppliers of essential items like motor oil or brake pads. . . .
>
> The obvious solution for a cash-tight operation is to triage vendor payments to ensure that absolutely essential items are always there. But in Detroit, no one inside the transportation department could direct payments to the most important vendors. A bureaucrat working miles away in City Hall, not responsible to the transportation department (and, frankly, not responsible to anyone we could identify), decided who got paid and who didn't. That meant vendors supplying noncritical items were often paid even as public buses were sidelined.[47]

Imagine a world in which 40 percent of the streetlights don't work. Almost one-third of the ambulances don't work, and many of those that do have over 250,000 miles on them. Some neighborhoods are so dangerous that ambulances won't enter them without a police escort. The average response time for the police is nearly an hour.[48] In the face of this public-safety crisis, the politicians cut the police force by 40 percent and closed most police stations to the public sixteen hours a day.[49]

Crime has soared as a result of these cuts. You are eleven times more likely to be killed in Detroit than you are in New York. The rate of violent crime in Motown is five times the national average, and the police solve fewer than 10 percent of the crimes committed there.[50] (You thought the comparison to *Mad Max* was an exaggeration?)

Detroit's greatest challenge today isn't structural; it's human. As Mark Steyn reported in *National Review*, "Forty-seven percent of adults are functionally illiterate, which is about the same rate as the Central African Republic, which at least has the excuse that it was ruled throughout the Seventies by a cannibal emperor.... The illiterates include a recent president of the school board, Otis Mathis, which doesn't bode well for the potential work force a decade hence."[51]

Detroit is not unique. Its tragic collapse is a warning of what will happen in the rest of the country if we continue to tolerate the massive, systemic breakdown of government. For two generations our political system has been dedicated to protecting the government class and growing a dependency class. Government employee unions, with their formidable electoral power, extracted more and more unsustainable deals from compliant politicians. And more and more citizens were told they didn't have to learn or work or be productive, because someone else would take care of it all for them.

The very complexity of this problem has paralyzed Congress. It is becoming more obvious to the American people with each passing year that the government is simply incapable of doing all the tasks it has assumed to itself. The result is a threefold breakdown of government.

There is a breakdown in simple competence. There is a breakdown in common sense and defined purpose. And there is a breakdown in the rule of law.

Let's look at each one in turn.

The Breakdown of Competence

Large elements of the government simply do not work anymore. Sometimes the cause is the personal incompetence of people protected by a civil service system in which firing the incompetent is very, very hard. (*USA Today* found that in many federal agencies, "death—rather than poor performance, misconduct or layoffs—is the primary threat to job security.")[52] Sometimes it is the attitude of people who have gotten used to their jobs and wonder why you are bothering them. Sometimes it is the complexity of bureaucratic systems that inhibit competence. Sometimes it is arcane regulations that, individually, might have been adopted for good reasons but collectively make it almost impossible to act.

The *Washington Post*'s David Ignatius captured the systemic incompetence in a column on the Obama administration's disingenuous public statements about the Benghazi attacks of September 11, 2012. When you look through the record of emails, "what you find is a 100-page novella of turf-battling and backside-covering," he wrote. "By the end, the original product is so shredded and pre-chewed that it has lost most of its meaning. All the relevant agencies have had their say, and there's little left for the public.... [T]he cascade of bureaucratic logrolling and pettifoggery begins, as each new agency is called to the trough."[53]

Ignatius describes a system in which the whole is substantially less than the sum of its parts. It is an anti-team. Instead of each person's working with the others to accomplish more than he could have done alone, each individual works against the others, and they all accomplish less than they would have done alone. This malady afflicts most parts of the government.

I once advised a successful entrepreneur who had proposed establishing a new office for modernizing the federal government. He was then asked by the president to come into the government to bring the entrepreneurial spirit to that new office. On his first day, the general counsel for the department visited him, explaining for hours what he could not do and what the penalties would be if he violated the various rules and laws. By the end of his first day in the federal bureaucracy, my friend knew he had been immersed in an anti-innovation system that would make it difficult to get anything done.

This man's experience was not uncommon. It is the norm.

We have had decades of reports on incompetence of this sort. Inspector general after inspector general has warned about the tens of billions lost annually to waste and fraud. Nothing happens. Jim Frogue, my former colleague at the Center for Health Transformation, wrote a book called *Stop Paying the Crooks*, identifying between $70 billion and $110 billion every year of Medicare and Medicaid fraud (enough to meet the goal President Obama proposed a few years ago of cutting $1 trillion over ten years, without touching a single honest person). Nothing has happened. An IRS inspector general report found that more than 20 percent of the agency's earned income tax credit payments in 2012 were improper.[54] That is an estimated $11 billion to $13 billion last year in just one program. This problem has gone on for years.

The only striking improvement in simple competence of government that I have found in recent decades was the adoption of CompStat (computer statistics) by the New York City police force under Mayor Rudy Giuliani and Chief William Bratton in 1993. Their methodical, systematic insistence on competence led to a dramatic decline in crime and a new spirit of safety and enthusiasm in our largest city. In the process, three out of every four precinct captains either retired or were reassigned.

Until we are prepared to define expected outcomes, measure results, and either retrain or replace people who are currently incompetent or unwilling to do their jobs, government will never achieve even the minimum standard of simple competence.

The Breakdown of Common Sense and Defined Purpose

In 2013 the *Washington Post* told the story of Marty Hahne, a magician from Missouri. As part of his act, Marty pulls a rabbit out of a hat. "To do that," the *Post* reported, "Hahne has an official U.S. government license. Not for the magic. For the rabbit." In the summer of 2013, Marty got a letter from the U.S. Department of Agriculture ("Dear Members of Our Regulated Community") about that rabbit license. Pursuant to a forty-year-old law intended to regulate zoos and circuses, the feds demanded to see his "disaster plan" for the animal. They told him that if he didn't submit a written plan for protecting the rabbit in a fire, a flood, a tornado, a power outage or air conditioning failure, a dam break, or an ice storm (among other imaginable threats to leporine well-being), he could lose his rabbit license.[55] The *Post* published Marty's rabbit disaster plan online and highlighted the absurdities, including its promise to fulfill the rabbit's right to "exercise" in the event of an evacuation.

Around the same time, the *Post* reported on the troubles of a California raisin farmer who owes the federal government $650,000 in fines and 1.2 million pounds of raisins for refusing to hand over large portions of his annual harvest to the national raisin reserve, "a farm program created to solve a problem during the Truman administration, and never turned off."[56]

Much of modern bureaucracy developed in response to favoritism and corruption. If you wanted uniform rules applied professionally and impartially, the thinking went, you needed a bureaucracy.

But bureaucracies are not impersonal, mathematical systems operating with geometric precision. They are organisms that grow and evolve over time.

Seeking to exercise and expand its power, the federal bureaucracy has produced more and more rules of greater and greater specificity, often disconnected from common sense and the real world. These rules give bureaucrats a tremendous amount of discretionary power. They can always find ways that individuals, businesses, and organizations are not in compliance with their regulations.

Former congresswoman Nancy Johnson of Connecticut used to describe how inspectors came to the nursing homes in her district to measure the distance of fire extinguishers from the floor. A few inches too high or too low, and the nursing home would be fined.

This type of behavior occurs naturally within bureaucracies. Most agencies recruit people who are interested in the fields they regulate. For example, the Environmental Protection Agency tends to hire people who are passionate about the environment. New bureaucrats then learn the rules and culture of the agency. Those who support the goals and principles of the environmentalists who run the EPA get promoted. Those who raise commonsense questions about cost and practicality, on the other hand, find their careers sidetracked.

The key bureaucrats in most federal agencies live in the Washington area. Many of them have worked in one of those large downtown office buildings for twenty or thirty years. They get their information from the Washington media. They often carpool to work together. Their children may go to the same schools. They have lunch together in the agency cafeteria. Some of them even vacation together.

Over time, these officials come to believe they are more important than the "ignorant" and "uninformed" people outside Washington. But in fact, the opposite is true. The Washington bureaucrats are too insulated to understand how their regulations work in the real world.

It was bureaucrats like these at the EPA who seriously considered regulating dust in rural America. A paper-pusher in an air-conditioned Washington office building can come to think that a regulation on dust makes sense. A farmer or a rancher knows it's insane. At a meeting in Arizona, one man said to me, "Do these bureaucrats understand Arizona is largely desert? Do they have any idea why we call one of our weather phenomena 'dust storms'?" The whole group roared with laughter.

In Iowa, farmers were simply incredulous that any bureaucracy could propose the kind of rules they were hearing about. They pointed out that dust routinely went back and forth and was irrelevant. As practical farmers, they didn't care. The idea of turning dust from a practical reality into pollution struck them as a sign of Washington insanity. Fortunately, their senator Chuck Grassley got word of the dust regulation and was able to derail it.

We do not have to tolerate the growth of a fourth branch of government called the bureaucracy with powers in many ways greater than those of Congress or the president. Lord Acton's warning that "power tends to corrupt, and absolute power corrupts absolutely" applies to bureaucrats as well as to elected officials. Their power has to be reviewed, trimmed, and at times eliminated.

Finding a commonsense balance between a lack of rules, which leads to cronyism and corruption, and an overabundance of rigid rules, which leads to the same, is an urgent task for citizens and elected officials.

As great a danger as regulatory overreach is, however, it is not the most perilous breakdown we face.

The Breakdown of the Rule of Law

The most frightening breakdown in government in our generation is the breakdown of the rule of law.

America was founded on the rule of law. The Founding Fathers understood that only the rule of law guarantees justice and opportunity for every citizen. They regarded the king's violation of the rule of law as the chief justification for the American Revolution.

The rule of law is essential to freedom because it's what stands between us and a capricious government that rewards its friends and hurts its enemies. It protects the weak from the powerful, the minority from the majority, the poor from the wealthy, and the little guy from the insider who can call on his contacts for special treatment. The rule of law provides the framework of certainty and fairness in which we plan our lives and our economic activities. It makes freedom and free enterprise possible.

The sense that we are protected by the rule of law, that Americans of every background enjoy a fair playing field with fair rules, has run deep in our civic culture for most of our history. It has been crucial to our success.

Indeed, in 1974, the commitment to the rule of law was deep enough that President Richard Nixon was forced out of office just two years after winning one of the biggest electoral victories in history. Americans understood then that using the Internal Revenue Service to go after your opponents is a fundamental violation of the rule of law. Using the Federal Bureau of Investigation to hide rather than to investigate crimes and lying under oath threaten the very fabric of the law because they prevent the people from getting to the truth.

Forty years later our culture tolerates or endures violations that would have been considered outrageous, impeachable offenses in Nixon's day.

Consider President Obama's Internal Revenue Service scandal. IRS officials targeted conservative, Tea Party, and patriotic groups, preventing them from getting the tax-exempt status they needed to raise money and demanding lists of their donors, the amount of each donation, and a host of details about their activities. Some liberal groups were slow to be approved, as well, skeptics have suggested. These left-wing groups, however, endured nothing close to the same level of scrutiny. The IRS officials investigated the Tea Party breakfasts hosted by an eighty-three-year-old grandmother who had been held in an internment camp during World War II, and they required that a pro-life organization divulge "the content of the members'...prayers."[57] In fact, not a single group identified with the Tea Party was approved for twenty-seven months after February 2010.[58] When caught, senior IRS officials lied about the behavior.

There are two possible explanations for using the IRS to harass conservative organizations. Both are chilling.

In the Watergate-style scenario, the president's top aides would encourage the Internal Revenue Service to go after the president's opponents. That would be illegal and dangerous, but it would at least have a clear cause-and-effect pattern. But the administration offers a different explanation: lower-level bureaucrats decided on their own that they would break the law and discriminate against conservative activists. That explanation is in some ways even more frightening. Has the rule of law deteriorated to the point that any bureaucrat can decide that his own values should define the law—that he can pick the winners and the losers?

James Bovard, writing in the *Wall Street Journal*, cites alarming evidence of a culture of corruption that has infected the IRS for some time: "A 1991 survey of 800 IRS executives and managers by the nonprofit Josephson Institute of Ethics revealed that three out of four respondents felt entitled to deceive or lie when testifying before a congressional committee."[59] The study is old, but has IRS culture changed? Recent reports suggest it has not.

Other examples of this breakdown have piled up in the last few years. EPA officials released personal information about thousands of farmers to environmental activist groups.[60] Similarly, the IRS released confidential information about conservative organizations to a competing liberal group.[61] The secretary of Health and Human Services, Kathleen Sebelius, has been shaking down private healthcare companies for funds to implement Obamacare after Congress denied her budget request.[62] The Justice Department has trampled the First Amendment by conducting criminal investigations of journalists who reported classified information, while the administration overlooked (and maybe even encouraged) serious national security leaks that burnished the president's image.

If the rule of law has indeed broken down so badly that every bureaucrat begins to think he can bend or break the rules to fit his own prejudices, then we are entering a dangerous world. And the Obama administration, it must be said, has led the way. The president has suspended parts of the duly enacted Obamacare law on his own. He has suspended the enforcement of immigration law on his own. He has asserted the authority to waive the requirements of welfare law on his own. In short, the president has undermined the fundamental principle of the rule of law.

Government is at the heart of the prison guards' ability to impose the status quo on the American people. Not all prison guards are part of the government, but virtually all rely on its power to preserve their privileges. Even as the old order uses government to defend the past, however, we are seeing a historic breakdown in the capacity of government to function. This has become a gigantic barrier to an American breakout.

The widespread breakdown in government is now obvious to Americans of both parties. The failures are becoming a problem of daily life. Citizens across the country are looking for some way to break out of this obsolete mess that is serving the people so poorly.

It is clear that minor tinkering will fail. We need extensive and extraordinary change to replace, not to reform, the broken-down parts of government. We need breakthroughs so large they can bring down the bureaucratic state, as light bulbs made candles obsolete.

Many of the pioneers we have just met are developing breakthroughs on that scale, breakthroughs that can make the failing bureaucracies in education, in health, in energy, and in transportation not only obsolete but irrelevant, by eliminating the problems that justify their existence. Part of the breakout for America will be a breakout from this dysfunctional bureaucracy that is holding us back.

CHAPTER NINE

BREAKOUT IN GOVERNMENT

The advanced breakdown in government that we're seeing today is something new, and it is forcing Americans across the political spectrum to recognize that our current approach to government is failing. Ever-expanding and unaccountable bureaucracy, conspicuous waste of taxpayers' money, and widespread corruption are a fundamentally wrong model. That's how old European monarchies treated their subjects, not how the self-governing citizens of the American republic are to be treated.

A French visitor to the United States famously remarked on Americans' new and inventive way of managing their affairs, noting how different it was from anything he had seen in the old world. "Americans of all ages, all conditions, all minds constantly unite," Alexis de Tocqueville wrote in 1835. "Not only do they have commercial and industrial associations in which all take part, but they also have a thousand other kinds: religious, moral, grave, futile, very general and very particular, immense and very small; Americans use associations to give fêtes, to found seminaries, to build inns, to raise churches, to distribute books, to send missionaries to the antipodes; in this manner they create hospitals, prisons, schools."[1] Citizenship, Tocqueville observed, was more than voting and complaining about the president. It was a novel way of life: do-it-yourself government.

Benjamin Franklin was the model of this new citizen-activist. He helped found the Junto, a group that met to discuss issues and ideas, when he was twenty-one. At twenty-five he helped found a library that was housed in Independence Hall for a number of years. At thirty he helped found a volunteer fire department. At thirty-seven Franklin proposed the Academy and College of Philadelphia. Over a six-year period, he helped launch the school, which became the University of Pennsylvania. The same year that he proposed the college, 1743, he also helped found the American Philosophical Society, which became the center of an astonishing range of scientific observations and inventions. In 1751, he helped found the first hospital in the American colonies. When you remember that Franklin was earning a good living, writing constantly, engaging in world-renowned scientific research, and playing a leading role in Pennsylvania and American politics and diplomacy, you can see why he has been called the "first American."

This is the ethos that produced the innovations and breakouts of earlier generations. It is an irony of history that the very modernity that the decentralized American culture made possible soon became a threat to that culture. In the nineteenth century, intellectuals became confident that they could understand economic and social forces well enough to manipulate them, much as they thought they had mastered the physical sciences.

The new complexity of the world seemed to demand centralization and professionalization. In his second inaugural address, President Franklin Roosevelt theorized that "as intricacies of human relationships increase, so power to govern them also must increase."[2] Bureaucracy was the answer to that challenge, a modern marvel—a vast administrative machine that could fine-tune society with the tools of science.

The new model of a centralized administrative state managed by a professional bureaucracy appealed to intellectuals eager to exercise enlightened authority over their fellow citizens. These intellectual Brahmins—a caste epitomized by FDR's celebrated "brain trust"—would be the architects and governors of a brave new world based on their values and their interests.

Roosevelt's New Deal created thirty new federal agencies for this task.[3] Foremost among them was the National Recovery Administration, or NRA, which had sweeping power to direct the economy. As Amity Shlaes recounts in *The Forgotten Man*, "NRA code determined the precise components of macaroni; it determined what tailors could and could not sew. In the poultry industry the...code had barred consumers from picking their own chickens. Customers had to take the run of the coop, a rule known as 'straight killing.' The idea was to increase efficiency."[4]

Eventually the Supreme Court found the NRA itself unconstitutional, but the expansive new regulatory bureaucracy remained, growing throughout the twentieth century regardless of who was in the White House or who controlled Congress. The public servants became the public masters. That is the legacy we are stuck with today—a central bureaucracy built in the 1930s trying to control a country of 315 million citizens. You might as well try to run the internet through a 1950s mainframe computer. The apparatus that once seemed scientific and modern is broken down beyond usefulness.

Computing overcame the limitations of cumbersome mainframes by distributing power among thousands of smaller computers in a decentralized network—the internet. That same technology can enable us for the first time to overcome the failure of centralized bureaucracy by replacing it with a network of connected citizens. Distributed government in the iPhone age should permit citizens to work together, voluntarily, on a scale that has never been possible before.

Citizenville

Gavin Newsom isn't the type of person you expect to find theorizing about how to dismantle vast government bureaucracies. A Democrat and the lieutenant governor of California, Newsom is the second-highest elected official in one of the nation's bluest states—one famous for its commitment to "big government" and for its particularly powerful public employee unions. (It was in California, after all, that the prison guard was making more than $200,000 a year.)

There's no question that Newsom is firmly on the left of the political spectrum. In his former office, mayor of San Francisco, he became one of the first officials in the country to issue marriage licenses to same-sex couples back in 2004. Later, with more controversy, he made San Francisco a sanctuary city from federal immigration law.

But Newsom has also had some experience dealing with bureaucracy, both as a small businessman and then as a public official. As lieutenant governor, he has been outspoken about the need to rethink government for the digital age. His book *Citizenville* is full of ideas about how Americans could use technology to break out of the outmoded bureaucracy that built the sidewalks to nowhere in Oklahoma.

"Citizenville" is Newsom's play on the title of the wildly popular Facebook game FarmVille, in which players tend a virtual farm, plowing land, planting and watering crops, and raising animals, for which they are awarded points. This might not sound like the most stimulating activity, but apparently a lot of people find it entertaining. FarmVille has been one of the top games on Facebook for years, with tens of millions of people spending hours a week tending to their digital fields. According to the game's publisher, Zynga, forty million people in more than 180 countries play FarmVille each month. Every day, thirty million people "visit" with fellow farmers within the game. In Turkey, one out of eight persons is on the game. Many players even buy virtual ornaments and accessories for their virtual farms using real money. Someone in Denmark purchased 3,700 pink flamingos at once to decorate what must be the most colorful agricultural scene ever envisioned.[5]

With so many people spending so much time and even money playing FarmVille to acquire worthless points, Newsom wondered if we could direct that energy into something more useful. What if we could turn government—or at least some important parts of it—into a game? What if players, instead of competing over harvests in FarmVille, could compete over real contributions in Citizenville? Newsom describes how it might work: "The way to 'win' Citizenville is to amass points by doing real-life good. If a player contacts the city to report a pothole and get it fixed, he gets one hundred points. If another player organizes a community cleanup in the local park, she gets two hundred points. If another player landscapes the median on his street, that's three hundred points. Whenever people make a real-life improvement, they report it to the Citizenville Website, which has a continuously updating scoreboard."[6]

If you're less optimistic than Newsom that Americans would compete over something as mundane as civic-mindedness, recall that Salman Khan has millions of kids frantically working for virtual badges in mathematics on his website. Compared with awards for dividing fractions, or for that matter for tending a virtual vegetable patch, getting rewarded for snapping a picture of an undiscovered pothole could be downright exciting.

Citizenville, however, should be able to do more than collect pictures of potholes for government employees to fill. Although Newsom doesn't go so far as to say it himself, it's not hard to imagine a version 2.0 in which players could capture the pothole on their phones and upload it to the Citizenville app, which could immediately shoot out a request for bids to local contractors and automatically select the highest-rated one. They could fill the pothole by the end of the day, a few citizens could verify, and the contractor could be paid automatically. A similar process might apply to any number of public services in Citizenville, from plowing snow off the roads to maintaining the community basketball courts.

Newsom even theorizes that Citizenville could re-create in the real world some of the magic that led that Dane to purchase those 3,700 flamingos. "In Citizenville," Newsom writes, "people would spend money on actual improvements in the player's neighborhood—say, an hour of professional landscaping or fresh paint to cover up graffiti. In both FarmVille and Citizenville, players have the enjoyment of the game. But in Citizenville, instead of taking pride in a virtual world, players would be making a difference in their own neighborhoods."[7]

Is all of this too crazy to work? Absolutely not—in fact, it's already happening. The prototype of Citizenville, or maybe its forerunner, has been up and running for four years—and not in spacey Silicon Valley, but in a tiny suburb of Austin, Texas. The city of Manor is pushing the frontiers of what government can be in the internet age. As Newsom describes, in 2009 the city launched a web-based platform called Manor Labs, which invited citizens to propose solutions to problems of local government. Instead of virtual points like in FarmVille, the program rewards participants in "innobucks"—one thousand for each suggestion, plus a bonus of one hundred thousand if the city actually adopts it. An online scoreboard keeps track of which citizens have acquired the most.

With suggestions rolling in, Newsom reports, the town took the experiment one step further: "The City of Manor came up with real rewards you could buy with your innobucks. For varying amounts, you could buy a police ride-along or even be mayor for the day. Local businesses and restaurants also got in on the fun, offering coupons for discounts or free appetizers in exchange for innobucks. It's not fake currency—it's civic currency."[8]

The "game" was a hit, building a group of active citizens who were addicted to improving local government. "When people went away on vacation," Newsom says, "they'd immediately interact with city government upon returning, trying to make up for lost time and build up their innobucks stashes."[9] Some of their suggestions really did make government more efficient and better for the citizens, such as automatic payments on utility bills.

A number of other cities have adopted a platform called SeeClickFix. Using a website or smartphone apps, citizens of Chicago, Houston, Richmond, Albuquerque, and other cities report problems to their local governments. On a recent day in Houston, city officials responded to reports on SeeClickFix of a traffic light outage, a missing stop sign, and a broken manhole cover, acknowledging them all within minutes.

Manor Labs and SeeClickFix—and, for that matter, Citizenville—are just the earliest hints of how the internet and ubiquitous smartphones could transform government. Even the most basic public functions that we didn't think offered much room for improvement, the "meat and potatoes" like road maintenance and trash removal, are open to whole new frontiers.

As mayor of San Francisco, Newsom was aggressive about bringing the ethos of a California entrepreneur to government. The city started taking suggestions from citizens through Twitter. Then it launched "Open

311," a computer protocol for submitting complaints to government. It created an iPhone app, EcoFinder, to help residents find the closest recycling locations. Newsom opened the city up for "hackathons," marathons of computer coding in which people created public tools (apps) for, say, coordinating carpools or reporting the GPS location of city buses. He distributed handheld devices that let people vote on budgeting decisions from the comfort of their living rooms.

Newsom has also stockpiled dozens of other great ideas for using technology to return government to the citizens. He cites House Majority Leader Eric Cantor's YouCut, which lets citizens vote on which spending cuts Cantor brings to the floor. The project drew between five hundred thousand and a million votes a week. The majority leader's office launched Citizen Cosponsors in 2012, which lets Americans endorse and follow legislation they care about using Facebook. There are now more than three thousand bills available to cosponsor, covering everything from foreign aid to Egypt to requiring members of Congress to read legislation before passing it.

Newsom sees the potential of this technology to disrupt the administrative state right out of business. In words any Republican would applaud, he writes about using the connective power of the internet to revive the ethos that Tocqueville detected in America. "We need to allow people to bypass government," he says. "We must encourage them to take matters into their own hands, to look to themselves for solving problems rather than asking the government to do things for them." In another passage, Newsom quotes the most famous Republican about "thinking anew": "We have to disenthrall ourselves, as Abraham Lincoln used to say, of the notion that politicians and government institutions will solve our problems. The reality is, we have to be prepared to solve our own problems."[10]

The bureaucracy, Newsom recognizes, is obsolete. We would do well to replace it with a twenty-first-century upgrade: "We have managers of managers, supervisors of supervisors, and enough committees, subcommittees, groups, and subgroups to make a bureaucrat's head spin...That's completely unnecessary in this technological age. Our government is clogged with a dense layer of bureaucracy, a holdover from an earlier era that adds bloat and expense. It's like a clay layer, a filler that serves only to slow everything down. But technology can get rid of that clay layer by making it possible for people to bypass the usual bureaucratic morass."

He puts it delicately, but it's hard to miss that Newsom is talking about a dramatic downsizing of government bureaucracy: "Technology allows us to disintermediate," he writes. "To disintermediate just means *to get rid of the middleman....* I'm talking about people organizing themselves to solve problems, rather than complaining that the government isn't doing it."[11] What conservative could put it better?

Newsom has no illusions about how the bureaucrats will respond to the innovations he's proposing. He sees that they will try to kill the future if we let them have the chance. He agrees with Joe Trippi, Howard Dean's former campaign manager, that "[b]ureaucracy wants to stop innovation," adding that it's "slow to adapt at best. At worst, it's openly hostile to change." (Of course, the solution to that willful resistance would be simply to defund it.)

A government that returned power from Washington to connected citizens would put a stop to much of the idiocy that Senator Coburn cataloged in his "wastebooks." After all, citizens engaged in local self-government are unlikely to spend their money on sidewalks to nowhere, robotic squirrels, or caviar advertisements. Nor do they sit around thinking up how to regulate themselves silly.

Of course, technology can't replace every government agency. The executive branch isn't going to disappear. But Newsom also has some fascinating thoughts about how to introduce competition into the functions of public administration that citizens can't take on themselves. One idea is to set up "an intracity competition among departments, with a Yelp-style scoreboard showing who has the highest rating." A public system for reviewing and scoring agencies, giving citizens a forum to praise good service and criticize bad or apathetic treatment, would go a long way toward improving departments of motor vehicles, building permit offices, and even courts. Newsom thinks it would fire the competitive instincts of public employees, but the awareness that bad interactions could quickly become public knowledge would also foster better behavior. There's already some evidence that it does. When Rialto, California, started requiring some police officers at all times to wear video cameras that kept a record of their actions, there was an 88-percent drop in the number of complaints filed against officers in one year.[12]

The truly revolutionary uses of technology aren't possible unless citizen computer whizzes can get the data they need out of government and into their apps. As with Facebook, Google Maps, or WebMD, the data are everything. Easy, real-time access to detailed information about government expenditures, for instance, would make it much more difficult for Congress or the bureaucracy to throw millions away on pancake restaurants and pet shampoo. Senator Coburn employed a staff spending hundreds of hours compiling that information. But if Amazon can follow you around the internet to identify your tastes and Visa can notify you instantly if someone uses your card to buy a television at a Best Buy across the country, transparency in federal spending should be a matter of routine.

Unsurprisingly, bureaucrats are not always eager to make data about their performance and spending accessible on every iPad. To its credit, the Obama administration has been moving in the right direction on this issue: the federal government does have a number of initiatives to open up federal data, but the sets published on data.gov often serve little purpose, like maps of clean energy companies, charts of electricity prices, satellite images, and census data. They are useful, perhaps—maybe even necessary—for Citizenville to become a reality, but they have little to do with holding government accountable. The useful data are often overly aggregated, or released in file formats that are useless for analysis (like PDFs). Typically Americans have to file Freedom of Information Act requests to get the kind of detailed information that would make citizen oversight possible.

Government's first step into the digital age is ensuring that state and federal agencies release all their data in standardized, machine-readable formats. In 1995, when I was Speaker of the House and only a few Americans had internet access, we created THOMAS at the Library of Congress to track legislation and make it available for the public to read online, along with voting records, sponsors, and other relevant information.

More recently, Representative Darrell Issa, the chairman of the House Committee on Oversight and Government Reform, has championed the Digital Accountability and Transparency Act (DATA), which would improve the information that bureaucracies submit to the federal spending database at usaspending.gov. The Democratic-controlled Senate has killed the measure once already.[13]

Tim O'Reilly, who publishes a popular series of handbooks for computer programmers, thinks the next big step is turning that one-way stream of data into a two-way dialogue. He coined the term "Web 2.0" to refer to a new kind of web services in the mid-2000s, sites like YouTube, Facebook, and Wikipedia. It's hard to remember now, but these websites were radically different from earlier internet sites. Instead of simply downloading pages from the web, as when you read an article at NYTimes.com, for instance, users were suddenly uploading information of their own as well. Contributing to Wikipedia, posting on Facebook, and adding videos to YouTube made the web a two-way street. The internet was taken over by its users.

In 2009, O'Reilly was perhaps the first person to suggest that the internet could bring the same transformation to government. He called his vision "Gov 2.0." Instead of sucking down whatever programs and services government provided, he thought, the users—the citizens—could make government a two-way street as well. They could take it back.

"The key idea is this," O'Reilly explained to Gavin Newsom, whose *Citizenville* is an imaginative example of what O'Reilly has in mind. "The best way for government to operate is to figure out what kinds of things are enablers of society and then make investments in those things. The same way that Apple figured out, 'If we turn the iPhone into a platform, outside developers will bring hundreds of thousands of applications to the table.' Previous smartphone development platforms looked like government does now: vendors talking in a back room and deciding what features to offer. But then Apple turned the iPhone into a platform in which the killer feature was that other people can make features."[14]

O'Reilly thinks that by treating government as a platform, we could replace many of the top-down functions of the bureaucracy with bottom-up solutions created by citizens. To do so, we have to let citizens compete directly with the lumbering state. "What if there were a market for government services, or government-like services," he asked Newsom, "where people could say, 'Oh, I can actually meet that need!' and there was a government apps store that wasn't 'Here are a bunch of apps the government developed' but 'Hey, we're gonna let you figure out how to compete with the government'?"[15]

This will mean more than opening up government data to the public, although that is an important step. O'Reilly thinks the government needs to create APIs (the technology that lets your iPhone apps give commands to, say, Twitter, on your behalf). Government needs to enable citizen apps to "plug in" to its services so that they can, for example, process forms, obtain permits, or pay taxes. A Citizenville app, for instance, would need to interface with local government to submit the pothole report, to authorize a contractor to fill it, and to pay them when the job was done. The government might have systems it uses for each of those tasks. O'Reilly wants to give programmers access to them as well.

With enough capabilities, citizen-created apps could begin to manage government better than the bureaucracies do themselves—not a high hurdle by any stretch. What would Kickstarter for government look like, to let citizens take direct control of a portion of the budget? What would Google Now for government look like, or Groupon, or eBay, or Airbnb?

The main barrier to this computer-driven decentralization of government isn't the technology, it's the policy. Government has to open up. O'Reilly gives the example of GPS, the now-ubiquitous location technology. "It's easy to forget that GPS, like the original internet, is a service kickstarted by the government," he says. "Here's the key point: the Air Force originally launched GPS satellites for its own purposes, but in a crucial *policy* decision, agreed to release a less accurate signal for commercial use. The Air Force moved from providing an application to providing a platform, with the result being a wave of innovation in the private sector."[16] When we're talking about a platform that could transform government and shrink the massive federal bureaucracy, however, there's little doubt that citizens will have to force this change themselves. The prison guards aren't simply going to surrender.

Both Newsom and O'Reilly seem optimistic, and they both—somewhat unexpectedly—seem excited by the promise that technology will carry America not just into the future, but into a future closer to the ideals of our founders and to the vibrant civil society Tocqueville so admired.

"The more actively engaged citizens are, the closer we come to the original vision of the Founding Fathers.... Aren't these the principles our country was built on?" Newsom asks. "Throughout the country, people took an active part, whether formally or informally, in the running of their towns—a kind of bottom-up democracy that served to strengthen the new nation."

O'Reilly echoed a similar sentiment. "If you go back to the Founding Fathers, you realize, 'Oh yeah, government was very small.' You go back to volunteer fire departments, to mutual insurance companies, to subscription libraries," he said. "But now we've created—government is like this special category of thing.... As opposed to, it's a set of things we do for each other."

"We've gotten very far away from that notion in recent decades," Newsom concludes, "a trend we'd do well to reverse."

Taming the Bureaucracy

Gavin Newsom is right. We *have* gotten very far away from that notion in recent decades. To give you some idea of how far and how fast: in 1929, federal spending accounted for just over 3 percent of GDP, and by 2010, that figure had exploded to 24 percent—a sevenfold increase.[17] And reading Senator Coburn's wastebooks, we know exactly how it has grown and how much it wastes.

Much of Washington has forgotten that government is not the whole society. It is just a small piece of it—or at least it should be. What communities and citizen volunteers can do, government should leave to them to do. What state and local governments can do, the federal government should let them do. That's the only way government can be dynamic enough to give ideas like Citizenville a chance to compete and flourish.

To shift responsibility from Washington to the states and local communities will require real political action. One of the most important policy proposals of my presidential campaign was a Tenth Amendment enforcement act. The Tenth Amendment provides that any powers not explicitly granted to the federal government under the Constitution "are reserved to the States respectively, or to the people." Yes, you read that correctly. The bureaucratic and regulatory monstrosity that we call the federal government is impossible to reconcile with that principle, which is why the prison guards want you to forget all about the Tenth Amendment. We the people will have a fight on our hands to bring about a genuine breakout from the failing government that holds us prisoner. While we're hacking away at the bureaucracy, we have to grow citizens at home. This is how Citizenville, Gov 2.0, and similar ideas will replace the bureaucrats that fill all those buildings along the National Mall.

Restoring the constitutional balance prescribed by the Tenth Amendment doesn't mean getting rid of the federal government entirely—not by a long shot. The Defense Department, the State Department, the Justice Department, and many others serve valuable public functions. But while they're important organs of government, many are also broken-down bureaucracies in critical need of reform. They're wasteful, inefficient, and run largely by an unaccountable army of federal employees. The 130-year-old civil service system was originally intended to stamp out patronage and corruption. But the reformed system developed corruptions of its own: it has organized to protect itself from the public and remain impervious to change, and it is staffed by people who can't be fired and are constantly growing their power and their pay. There is no incentive for the current civil service to run the government efficiently or to save the taxpayers' money.

The bureaucrats behave accordingly. Their focus is entirely on *inputs* rather than *outcomes*. But inputs—more spending, more employees, more programs—are meaningless unless they produce better outcomes. And as we pour more money into the old bureaucracy, we get worse outcomes.

A meaningful program of civil service reform would impose modern technology and management practices on the federal bureaucracy. For instance, the federal government still uses hundreds of siloed computer systems at a time when many businesses have migrated to cheaper cloud-based solutions that serve tens of millions of people efficiently.

Lean Six Sigma

The great improvement in American productivity over the last half century was brought about by the application of W. Edwards Deming's vision of quality through continuous improvement and more recently by the development of management systems and philosophies like the "Lean" system for process improvement.

In Iowa as a candidate for president, I encountered Mike George, a remarkable champion of this type of modernization for government. One summer morning in Des Moines, his organization drew nearly a thousand concerned citizens to its Deficit Free America Summit to learn about his unusual plan to balance the federal budget by 2017—a bold idea, since many in Washington claim that it will take at least a few decades to balance the budget.

Mike knew from personal experience that we could achieve massive cost reductions (or lower inputs) while dramatically improving outcomes.

As a business consultant, Mike had helped create the Lean Six Sigma method for deep cost cuts, and through his firm, the George Group, he had brought his expertise at streamlining corporate processes to companies like Xerox and Caterpillar. Using his method, Motorola was able to return 1,500 manufacturing jobs to the United States that it had previously outsourced to China.

Mike's Lean Six Sigma method was so successful at cutting waste in private enterprise that in 2004 the Department of Defense hired the George Group to tame runaway costs by training Pentagon personnel in the process. The results were stunning.

The army reduced costs by approximately $22 billion in the processes and programs to which it applied Lean Six Sigma. The Army Materiel Command alone is responsible for half of that savings by removing waste from the army's supply chain. The Naval Warfare Systems Center in Charleston, South Carolina, increased the production of mine-resistant, ambush-protected vehicles (used to protect soldiers from roadside bombs) by a factor of ten, with no new facilities and no additional employees. Another navy facility achieved a fivefold increase in productivity and an 83-percent reduction in cycle time on aircraft engine repairs.

The army has even established annual awards for excellence in Lean Six Sigma. Its goal is to have 0.5 percent of the army workforce certified in the method to ensure that resources do not go to waste and service members get the support they deserve.

One member of Congress who understands this concept is Chris Collins of Buffalo, New York. When he was the county executive of Erie County, he implemented Lean Six Sigma in a heavily unionized workforce. He streamlined hiring for county jobs, improved the collection of delinquent taxes, used vehicle fleets more efficiently, and even cut the number of copy machines in county offices. The savings added up to almost $12 million in just a few years. Lean Six Sigma produced better services, a better climate for creating jobs, and lower costs for the taxpayers of Erie County.

Like Chris Collins, Mike George thought there was no good reason to accept continual increases in spending across the federal government without the kind of careful examination that has reduced the cost of many military projects. So after giving up his financial interest in Lean Six Sigma in 2007, he decided to get involved in solving our nation's fiscal challenges.

Mike estimates that roughly 25 percent of all government spending is waste. One example he cites is a National Institute of Medicine and National Academy of Engineering joint study that concluded that 30 to 40 percent of all healthcare spending (a major budget item) is waste. Eliminating the one-quarter of federal spending that is waste would slash between $500 and $700 billion per year from the federal budget. That amount is roughly equal to the entire annual budget of the Department of Defense. It's $5 trillion over ten years.

Cutting out the waste, however, will not get rid of the federal morass. Americans must also demand a complete rethinking of the regulatory state, which is killing innovation and strangling our economy. Every major regulatory agency should be reorganized, and every major regulation reassessed with the goal of removing the barriers to innovation.

A crucial step toward decentralized government will be limiting federal regulations to matters involving interstate commerce. Our Constitution requires this, actually, though you'd never infer it from the federal laws and regulations that reach into every corner of our lives. In most areas, states and local governments have a better sense of what rules are necessary for their communities. Distributing regulatory authority among the states also has the important advantage of letting them compete with each other to be efficient and friendly to innovation. Indeed, federal regulators should defer to states when they have corresponding regulatory agencies rather than overriding them with federal rules. Federal agencies should bear the burden of proof that their proposed regulations affect interstate commerce. If agencies stray beyond their constitutional authority and a regulation is struck down in court, they should be held responsible for the compliance costs that businesses have incurred trying to adjust to the improper exercise of power. Citizens should not suffer from irresponsible rule-making.

Federal regulations that do meet the Tenth-Amendment standard should be subject to a mandatory analysis detailing the effect on innovation, economic growth, and jobs. The cost of the 800-percent increase in regulation since 1949—eating away as much as three-quarters of economic growth—is too disastrous to ignore. Agencies like the EPA typically produce analyses of proposed regulations that consider "environmental justice," "human dignity," and other utterly subjective concerns with little regard for the adverse economic effects of their decrees. They should adjust their assessments of "human dignity" and "environmental justice" to take account of economic harm to the poor and disadvantaged when issuing new rules. And federal regulators should be forced to factor innovation and economic growth—the forces that really make our lives better—into all of their analysis.

CHAPTER TEN

BREAKOUT FROM POVERTY

No one in America needs breakout more than the poor.

The poorest Americans are the most firmly trapped prisoners in our country. The schools where they live are unimaginably bad, but they have no real alternatives. Once those schools have passed them through, they have no second chance to learn the skills they need to prosper in the modern world.

The cities they inhabit are broken down, driven to the verge of bankruptcy by mismanagement and corruption. Public safety and public services are appalling failures, all but killing opportunity for the people trapped there.

And for the last fifty years, a torrent of federal anti-poverty programs, developed with the best of intentions, have actually made things worse.

Growth

Unfortunately, the economic stagnation that has afflicted the United States since the financial crisis of 2008 has made life even more difficult for the poor. The poverty rate is now at its highest level since the 1960s.[1]

President John F. Kennedy was right: "A rising tide lifts all boats." But if the tide is receding, the rickety craft of the poor get beached first. Restoring economic growth is therefore a moral concern. Real economic growth does more to help the poor than any social program.

In 1993, nearly forty million Americans (15.1 percent of the population) lived in poverty.[2] Five of the next seven years saw economic growth over 4 percent.[3] By 2000, the number of Americans living in poverty had fallen to thirty-one million (11.3 percent of the population).[4] Thanks to a growing economy, nearly a quarter of all Americans who were living in poverty in 1993 had climbed out by 2000. According to the U.S. census, the poorest of the poor enjoyed the most improvement. Whether you sliced by age or ethnicity, "those groups with higher poverty rates had their rates fall further than those with lower poverty rates."[5]

Economic growth is not just a facile catchphrase. It is *the* best opportunity for the poor to climb out of poverty. And an economic breakout—spectacular growth fueled by dramatic technological leaps and favorable government policy—would be a catapult for the poor.

Economists have long understood that innovation is the primary engine of growth. Several years ago a senior research analyst at J.P. Morgan produced a graph of data from the late economic historian Angus Maddison illustrating the growth of the world's gross domestic product per capita over the past two thousand years. Until about 1800, all of humanity is clumped together in a flat line near the bottom of the graph. With the Industrial Revolution, Western Europe and the United States shoot sharply up, from a GDP per capita of a few hundred dollars to thousands. Suddenly, as if a light was turned on, they have modern, prosperous economies, and the average man is living better than anyone before in human history. Then with the innovations of the twentieth century, the lines representing Japan, the United States, and Western Europe go asymptotic—nearly vertical.[6]

One hundred percent of Americans live better and more prosperous lives because of the innovation and economic growth of the last half century. The economist Deirdre McCloskey estimates Americans today are so much richer on average than they were in 1900 that "if one accounts at their proper value novelties such as jet travel and vitamin pills and instant messaging, then the factor of material improvement climbs even higher than sixteen—to eighteen, or thirty, or far beyond." One century of innovation and economic growth made us perhaps thirty times richer.[7] Someday, economic historians could make similar calculations to consider how much self-driving cars, regenerated kidneys, and adaptive learning systems improved our lives and made all of us more prosperous.

Breakthrough innovations and the economic growth they stimulated are what separated Lincoln's era from Jefferson's, Edison's era from Lincoln's, and Einstein's era from Edison's. In each successive period, Americans were wealthier, more comfortable, and more mobile than before—even, perhaps especially, the poorest.

In the near future, big innovations like those I have outlined in this book could lead to a breakout for the American economy in general and for the poor in particular.

The breakout we're looking for depends on economic growth comparable to that of the Reagan recovery of the 1980s or the boom of the 1990s. But economic growth by itself is not enough. Many of the poor have benefited modestly from past growth in the general economy but remain trapped in poverty. We have to think carefully about why they have been left behind and what we can do to make sure every person in America has the opportunity to thrive.

The persons we statistically lump together on the basis of income and label "the poor" are actually enormously diverse. Some of the deeper patterns of poverty that differentiate them are familiar to most of us; others are less intuitive. In the miserable economic conditions since the 2008 crash, the ranks of the "working poor"—those who work long hours for low wages—have been growing. Rural and small-town Americans are left in poverty as the information and knowledge economy sends good jobs to larger cities. Many inner-city neighborhoods, meanwhile, are afflicted by crime and gangs and high unemployment. The tragically self-destructive cultures that have taken root there make their economic problems much worse.

The poverty rate among minorities is alarming. Thirty-eight percent of African American children live in poverty. Thirty-five percent of Latino children live in poverty. Native Americans living isolated on reservations with a different set of communal rules are another group with exceptionally high levels of poverty and unemployment.[8]

Immigrants often find their job options limited. Single mothers, too, are disproportionately likely to be poor. The extraordinary number of Americans under the watch of the corrections system in one form or another—especially prisoners and former prisoners—face challenges of their own. Americans with disabilities have a completely different set of problems keeping them from working (see chapter twelve for details). Alcoholism, drug addiction, and mental illness are major problems among the homeless and very poor.

The economy has been so bad in recent years that poverty of a different kind is touching a group that has done well historically. Educated young people are leaving college to face a bleak job market and burdensome student loan repayments.

Senior citizens are living longer and staying healthier than anyone expected, which is good news, but many of them rely on pension plans that overpromised. Our culture and the tax laws, meanwhile, have encouraged borrowing and discouraged savings, so many older Americans find themselves working at low-paying jobs, either full- or part-time, long after they had planned to retire.

Another group of middle-aged and slightly older Americans have educations and skills no longer adequate to sustain the incomes they have been used to. Many of their employers are downsizing or closing. They want to work but they have the wrong skills for the twenty-first century and no access to the education they need to become employable.

Obamacare, with its powerful incentives to hire people for fewer than thirty hours a week, is exacerbating all these problems. For the first time in history, we have a part-time economy for millions of Americans who would like to have full-time jobs with full benefits.

Notice how different these groups are. Each faces its own challenges. For this diverse range of Americans to leave poverty for prosperity, we need breakthroughs big enough to reach all of them. In particular, it will take breakthroughs in education, welfare policy, and the penal system to create a breakout from poverty.

Learning

Learning is at the heart of leaving poverty. "Give a man a fish and you feed him for a day. Teach a man to fish and you feed him for a lifetime."

Any strategy for breaking out of poverty has to start with helping the poor acquire skills and knowledge. This imperative applies to so many patterns of poverty (the inner-city poor, the out-of-work in rural America, the middle-aged Americans who have lost their jobs in this bad economy, new immigrants who struggle to learn English, prisoners—all of them, really) that it must be a top priority.

It is obvious that the current education system is failing the people who need it most. Our inner-city schools are scenes of mayhem. Ron Suskind's book *A Hope in the Unseen*, based on his Pulitzer Prize–winning series in the *Wall Street Journal*, chronicles the school days of Cedric Jennings, a young African American at Ballou High School in Washington, D.C. Cedric is remarkable because he dares to be a good student in an environment where academic success is dangerous to one's personal safety. The "silent majority at Ballou . . . are duck-and-run adolescents." The gossip in the halls isn't about who has a crush on whom but about "a boy shot recently during lunch period, another hacked with an ax, the girl gang member wounded in a knife fight with a female rival, the weekly fires set in lockers and bathrooms, and that unidentified body dumped a few weeks ago behind the parking lot." The school's lesson to the students: "distinctiveness can be dangerous, so it's best to develop an aptitude for not being noticed."

The honor roll is "pinned up like the manifest from a plane crash, the names of survivors." It lists the students with at least a B average. Of Ballou's 1,389 students, seventy-nine have made the honor roll: sixty-seven girls and twelve boys.

With heroic determination and effort and the encouragement of a great teacher and devoted mother (his father is incarcerated), Cedric sets his sights on top-tier colleges—and in his senior year of high school, he's accepted into an Ivy League university. When he arrives on campus in the fall, however, Cedric finds himself academically far behind most of his fellow students and struggling to adjust.

Cedric was one of the miraculous few who made it through the brutal gauntlet. In seventeen of our fifty largest cities, fewer than half the students graduate from high school.[9] In New York City, only 30 percent of students are proficient in math and only 26 percent in reading.[10] Only 21 percent of Chicago eighth-graders are proficient in reading.[11] The schools aren't the only party to blame for this failure, but they are prison guards blocking the change that could help other poor kids like Cedric break out of poverty.

Compounding the problem, the education system has no way to retrieve the older Americans whom it has already failed. Forty-seven percent of Detroit's adult population is illiterate.[12] What hope is there for them to get a good job in the information age, even if they can escape the ruins of what was once the most innovative city in America?

Better-educated adults also face the imperative to learn new skills when their old jobs go away and they have to change careers. Indeed, the nature of work has been changing in fundamental ways. Peter Drucker, who began writing about this change fifty years ago, has observed that the level of education needed to rise from a low-paying job to a higher-paying one is increasing. It was clear as early as the 1960s that we needed an entirely new approach to adult learning if large numbers of workers were not to become obsolete in their thirties and forties. Yet we still have no system of lifelong learning to help workers update their skills. There is an enormous gap between the learning system we need and the expensive, obsolete, inconvenient, and inefficient system of bureaucratic education we are currently stuck with. We have to establish the principle that learning never ends, that workers need to pursue part-time learning their entire lives.

The current education system also lets down new immigrants, whose opportunities are drastically limited if they don't become proficient in English.

Fortunately, pioneers like Salman Khan are preparing the breakout in learning that could smash one of the greatest barriers to a breakout from poverty. We have already seen how Khan's approach to education can save students who have fallen far behind.[13] Learning science and technology could put many poor Americans on a trajectory for success. Within a few years, every poor person in America who wants it could have access to Khan Academy and other innovative learning materials. No longer will they be prisoners of failing schools.

The smartphone is becoming so common that we take it for granted. Even many poor Americans have an iPhone or an Android device on which they can run apps, watch videos, and communicate with people around the world. And guess what? The sophisticated adaptive learning software that Bror Saxberg describes could run on these devices.

Smartphones can bridge the information gap and empower the poor with materials tailored for their needs and abilities. They can emphasize audio and video over print, and they can provide the short, practical building blocks of learning that Khan Academy has shown to be so effective. Apps like Duolingo already teach foreign languages on the iPad for free. We need Duolingo for English literacy.

The information gap—previously an unbridgeable chasm—has often separated the poor from the prosperous. Now we have the ability to bridge that gap with an inexpensive device you can hold in your hand. A thousand revolutions could not yield as much progress for the poor as the iPhone might do.

Indeed, we have not yet thought through the implications of the smartphone for public policy. The rapid spread of these sophisticated devices has the potential to solve many big problems at no cost to the government.

A Change in Attitude

There is no doubt that learning is essential to helping people leave poverty. But they need to learn about more than mathematics or Microsoft Excel. People need to learn the right attitudes and habits. In fact, these skills are probably more important to success than 80 percent of what students learn in school. Habituation to work and the right attitudes are one of the education system's primary goals.

The next great requirement for a breakout from poverty, then, is to change the destructive attitudes and habits that hold so many people back. We have seen the power of the right attitude in the story of Cedric Jennings. We will see it in chapter twelve in the story of Congresswoman Tammy Duckworth. The right attitude and habits can carry people a long way. As Marvin Olasky points out in *The Tragedy of American Compassion*, the principle of helping the poor learn sound habits was at the heart of every nineteenth-century reformer's agenda.

Last year I met one of the world's leading poverty fighters. I was shocked when he said to me, "When people tell me they're unemployed, I say, 'Well, then why haven't you created a job?'" Perhaps not everyone can be an entrepreneur, but with ninety-nine weeks of unemployment compensation, they could obtain an associate's degree (or, with inexpensive online options, a bachelor's or even a master's degree). Those two years are more than enough time to start a business of your own. But how many think about long-term unemployment this way?

We know that graduates of self-sufficiency programs that focus on job training and on skills like budgeting and saving are more likely to be employed and have higher incomes.[14] The same learning technology that teaches reading and math could one day teach these important life skills as well, helping people acquire the habits and attitudes they need to leave poverty. Sebastian Thrun's Udacity already has a free online course called "How to Build a Startup," walking would-be entrepreneurs through questions like "How do you make your money?" and "How do you get, keep, and grow customers?"

What if this training could be delivered in a highly personalized way through smartphone apps designed specifically to help the poor? The unemployed could learn how they might earn money, get a job, or start a business of their own. They could learn about saving money and creating a budget. In short, online programs might help the poor become self-sufficient, leaving poverty for opportunity. Such programs might even include online mentors, volunteers who have made it out of poverty themselves, like the "learning coaches" who help Khan Academy students through their lessons.

Some people living in dependency have hours of idle time each week. We could soon ask anyone receiving public assistance to complete a training program in employment or life skills. A system of continuous learning and assessment might eventually include automated incentives to help people work their way through the programs, perhaps small amounts of cash. If someone learning to read or to get a job received a small but immediate cash reward as she completed each step, her focus on learning would improve dramatically.

As a congressman from Georgia in 1990, I ran an experiment called Earning by Learning. We paid "at-risk" students two dollars for every book they read during the summer. Word got around, and each week a few more students joined the program. A fourth-grader named Stephanie Wynn read more books than any other student. The *Wall Street Journal* invited her to send in her story, which it published unaltered.[15]

"Last summer I read 83 books," she began. "I earned $166.00, $2 for every book.... I spent a lot of time this summer reading. If I hadn't read the books, I would have been bored. I do like to swim and watch TV sometimes. But the reading time was fun."

Stephanie got some of her friends to start reading. "I recommended a book to my friend Jeremiah," she wrote. "I let him take it home and read it. He said it was really good. He brought it back to me."

At the end of the summer, Stephanie's school recognized her for reading the most books in the program. "All the kids were very happy to get their money," she said. "With the money I earned I bought some new clothes and got a lot of Barbie stuff."

"The summer reading program is over. I am still reading. I am still reading because it is fun," she concluded. "I think it is a good idea to give kids money for reading books. It showed me that reading was fun. It also helped bring my mom and I closer together. We had fun reading together."

At one time, groups in seventeen states had picked up Earning by Learning. Consider how easy an e-reader app might make such a program. It could even verify that the children read the books.

The Welfare State Isn't Working

The Earning by Learning reading program was hard to sustain because its combination of personalization and financial incentives violated the ethos of the welfare bureaucracy. In order to achieve a breakout from poverty, we have to replace the welfare state with real opportunity.

For the last fifty years, government poverty programs have made things worse. With the best of intentions, the paternalistic, patronizing, dependency-inducing model of welfare has produced illiteracy, unemployability, and hopelessness. It has weakened the poor.

In the mid-twentieth century, the United States entered into a destructive social contract with the poor. If you are a mother and you kick the male out of the house, we will give you money. If you have more children outside of marriage, we will give you more money. If you are a student and you do badly in school, we will subsidize you. If your body is imperfect, we will send you a disability check for the rest of your life, provided you don't work full time. It is a soul-crushing deal. Practically everything we have done has made it harder to break out of poverty in America.

From the Moynihan report of 1965 to Charles Murray's *Losing Ground* in the 1980s to Marvin Olasky's *The Tragedy of American Compassion* a decade later, the case against dependency is devastating. The sad reduction of citizens to clients is exactly the opposite of the American ideal, and the prison guards of the poor make it expensive to stop being a client.

The Heritage Foundation estimates that over a trillion dollars are spent each year on what Peter Ferrara identifies as roughly two hundred federal programs designed to transfer resources to the poor. Taking a job can cost someone unemployment compensation or disability payments or earned income tax credit or Medicaid or public housing or food stamps or some combination of these and other subsidies.

Instead of a glide path from poverty to prosperity, the welfare state has built a cliff that the poor have to climb to reach an income level where becoming independent makes financial sense. The available support is a spiderweb of dependency that can make it almost unthinkable to leave the government's welfare system and become independent.

Total welfare benefits today provide more than a minimum-wage job in thirty-four states and the District of Columbia, according to a recent study by the Cato Institute. In seven states and D.C., they provide more than a job paying twenty dollars an hour, and in five other states, welfare provides more than a job paying fifteen dollars an hour. In Hawaii, the study found, "a person leaving welfare for work would have to earn more than $60,590 a year to be better off." It's no wonder so many Americans choose dependency when, according to Cato, "In ten states and the District of Columbia, welfare pays more than the entry-level salary for a teacher in that state" and "in 38 states and the District of Columbia, welfare is more generous than the average starting salary for a secretary."[16]

The "Great Society" legislation was passed half a century ago, and the results speak for themselves. We waged war on poverty, and poverty won. The welfare-state bureaucracy cut off the bottom fifth of America from economic opportunity and independence. (And that bottom fifth has almost become the bottom quarter under the weight of a bad economy and an inadequate adult education system.) Rising became too difficult and sliding backward as a client of the welfare bureaucracy became too easy.

The welfare industry, however, resists real change and denies any responsibility for the mess. Half the adults in Detroit can't read, but don't blame the bureaucracy that has spent trillions over the last generation. For decades, the prison guards have subsidized unemployability, and now these illiterate adults can't get a job in the information age. If neighborhoods collapse under the dysfunction and everyone capable of moving leaves, don't blame the unions that insist on trapping each successive generation in failing schools. The education system is so bad, the population so poor, and the city laws so burdensome that the only place to make money is in the underground economy. But don't blame the prison guards for the city's degeneration into crime and violence.

We are dealing with human beings, so achievement needs to be rewarded and effort needs to be reinforced. If we took a third of the amount we're spending on the poor and spent it on sensible incentives, we could begin the breakout from the welfare prison.

The perpetuation of material poverty isn't the only problem with the welfare state, however. It also wreaks havoc on marriage and the family. Robert Rector of the Heritage Foundation puts it bluntly: "Welfare... converts the low-income working husband from a necessary breadwinner into a net financial handicap. It transformed marriage from a legal institution designed to protect and nurture children into an institution that financially penalizes nearly all low-income parents who enter into it."

Predictably, these terrible incentives contributed to a breakdown in the American family. As Peter Ferrara notes, only 7 percent of children were born to unmarried parents at the outset of the War on Poverty. The figure today is many times that: nearly two in five children nationwide are born to parents who are not married to each other. The numbers are starkest among African Americans, who are disproportionately poor. When the War on Poverty began, well over half of African American children were born into two-parent families. Sadly, as the welfare programs that penalized marriage began to take effect, the rate plummeted. The portion of African American children born out of wedlock rose from 28 percent in 1965 to 49 percent in 1975. By 1990 it was 65 percent. Today it is about 70 percent. White families, too, have collapsed. In 1965 the portion of Caucasian children born out of wedlock was just one in twenty-five. Today, it is one in four. Among white Americans without a high school degree, it is one-half of their children.

This is a disaster for the poor, for it feeds the cycle of poverty. We know, says Ferrara, that poverty and having children out of wedlock are strongly linked. The poverty rate in 2010 for married couples with children was 8.8 percent, compared with 40.7 percent for unmarried mothers.[17] Among African Americans, the poverty rate is 12 percent for two-parent homes. Among homes led by unwed mothers, it is nearly 50 percent.

Children in one-parent homes are virtually sentenced to poverty. They are seven times more likely to become welfare recipients as adults, says Ferrara.

The *New York Times* columnist Charles M. Blow acknowledges the perverse incentives of the welfare state: "For the poorest Americans, there are marriage penalties built into many of our welfare programs," he writes. "As the Heritage Foundation has pointed out: 'Marriage penalties occur in many means-tested programs, such as food stamps, public housing, Medicaid, day care and Temporary Assistance to Needy Families. The welfare system should be overhauled to reduce such counterproductive incentives.'" When liberal columnists are quoting the Heritage Foundation favorably in the *New York Times*, you know the time has come to rethink our entire approach to helping poor Americans.[18]

Corrections Reform

Mass incarceration is decimating the poor. Blow identifies it as one of the destructive public policies that have undermined marriage and opportunity: "In the two decades preceding the Great Recession, the American prison population nearly tripled, according to the Pew Center on the States. And make no mistake: mass incarceration rips at the fabric of families and whole communities."[19]

We need a profound rethinking of our corrections system in general and our prisons in particular. Chuck Colson's courageous founding of Prison Fellowship marked a resurgence of serious conservative thought about prison reform, and I have been working for years with his successor, Pat Nolan, on rethinking incarceration. It is critical to the future of a large part of our population, and of the African American community in particular, that we correct policies that turn first-time offenders into hardened criminals. We can do better. We owe it to America to do better.

One out of every thirty-one Americans is under correctional supervision.[20] This is the highest rate of any developed country. One-quarter of the world's prisoners are incarcerated in the United States. The link to poverty is obvious.

There is a striking correlation between the collapse of education for the poor and incarceration. Seventy percent of prisoners rank in the lowest two levels of reading ability, according to the National Institute for Literacy.[21] The Annie E. Casey Foundation found that "ten to fifteen percent of children with serious reading problems will drop out of high school, and about half of youth with criminal records or with a history of substance abuse have reading problems."[22] Many studies have shown that prisoners who obtain a GED while incarcerated are substantially less likely to return to prison than those who do not.[23]

The picture that emerges is of a terrible nexus—poverty, dependency, a failing education system, a destructive culture in which most children are born to unwed mothers, crime, and finally, prison—each malady reinforcing the others.

The current system is a human, social, and financial disaster. We lock up far too many nonviolent people. Our prisons are holding facilities often controlled by the criminals. Too many first-time prisoners learn how to be a criminal rather than how to read. Our recidivism rate (people going back to jail) is tragically high, over 60 percent.[24] The corrections system is not correcting.

Breakthroughs in technology suggest three innovative ways of dealing with offenders. The first step is obvious: Every prisoner in America should spend several hours a day working through Khan Academy or another free online learning system that measures his progress. When he has finished his secondary education, he should move on to programs like Udacity and Kaplan and start working on college courses for free. He should simultaneously take online classes that teach him how to return to society—and stay there—as a productive citizen. All privileges in prison and evaluations for parole should be tied to cooperation and achievement in such a system.

Even a decade ago, providing this level of education to prisoners would have been prohibitively expensive, but today it is free. Taxpayers should not pay for people to sit in prison and do nothing when we know that if they learn something, they will be less likely to return. An online education will reassure potential employers that someone improved in prison. If we can make convicts pick up trash on the side of the highway, we can make them complete every lesson on Khan Academy.

Second, we should replace prison with fines and electronically monitored parole for many nonviolent offenders so they can continue to work, keep their families together, and avoid learning from hardened criminals. We can track people precisely with GPS and establish where they are permitted to go, and even (with video) what they are permitted to do. For many nonviolent offenders, this would be a more productive, humane, and cost-effective form of correction than prison. For those who do go to jail, we should use similar electronic controls to take the prisons back from the intractable convicts who dominate them.

Finally, we should give parole officers and prison personnel incentives to rehabilitate and educate offenders rather than merely warehouse them. Van Jones, my liberal *Crossfire* cohost on CNN, proposes a bonus for wardens whose former inmates get a job and do not return to prison.

We cannot simply write off 2.3 million Americans in prison. We need a breakout in our thinking about incarceration in America.

Safety, Housing, and Jobs

Individual improvement and a reform of the existing institutions are important steps toward a breakout from poverty. But public safety, housing, and jobs are essential parts of the picture. We need to achieve breakthroughs in all three areas.

No city can be economically successful if it does not maintain public safety, which is the first responsibility of government. Yet local governments across the country have neglected that responsibility as they have spent their money and resources on other concerns. There is no reason that American cities can't be safe places to live and work if our local governments are smart and determined. As mayor of New York, Rudy Giuliani proved with his police commissioner, William Bratton, that technology and the prudent management of resources could produce a remarkable improvement in safety. Their CompStat program achieved a 75-percent reduction in crime between 1993 and

2005, making New York the safest big city in the United States. Mayor Michael Bloomberg and Commissioner Raymond Kelly have continued that trajectory. Bratton repeated the achievement in Los Angeles. It is clear that violence on the scale of Chicago or Detroit is a function of bad management and leadership.

If large cities are to enjoy safety, the authorities must control gangs. The FBI estimates that there are 1.4 million gang members in the United States (an astonishing 40-percent increase from 2009).[25] These gangs, the FBI says, are responsible for "an average of 48 percent of violent crime in most jurisdictions, and much higher in others." The Chicago police estimate that 80 percent of the city's murders are gang-related. Defeating gangs while weaning young people away from crime with better opportunities should be a major goal of any program for safe neighborhoods.

Housing is one of life's essentials, of course, and owning property is a key element of citizenship. The late Jack Kemp spent much of his congressional career helping the poor develop "sweat equity" in public housing. The idea is that the occupants of public housing assume responsibility for its maintenance and ultimately acquire equity in their homes. The prison guards of the past who most fiercely opposed this strategy were the public-housing unions. During one debate on sweat equity for public housing, one member of Congress from an inner-city district said to me privately that he agreed with Kemp but that a "yes" vote would be too expensive personally. When I asked how a vote in favor could cost him, his answer was simple: "If I vote no, I have no primary, and I can go on vacation for several months," he said. "If I vote yes, the public housing employees union will run a candidate against me. I will win, but I will have to raise half a million dollars and campaign for three or four months. It isn't worth the cost."

At another point during the debates in the House, Congressman Barney Frank of Massachusetts objected that if we allowed poor people to acquire sweat equity in their homes, seven or eight years later, they could sell the homes for several hundred thousand dollars—and then they would no longer be poor![26]

A sweat-equity model for public-housing occupants would help the poor attain pride in their neighborhoods and enable them to invest in their own future. The public-housing unions will hate being replaced by cooperative or condo associations, but the improvement in outlook and wealth for the poor they are supposedly serving will be clear.

Finally, jobs must be at the heart of any strategy for helping people work their way out of poverty. Opportunity for the poor has been scarce in the stagnant Obama economy. When millions of high school and even college graduates are out of work, why should they believe they can get a job?

We have already discussed the most promising and important ways of promoting employment: economic growth, lifelong learning, and ending the welfare state's incentives not to work are all crucial steps. And yet millions of Americans are working every day for less than a decent living. They feel frozen in poverty, and that feeling is a disturbing departure from the historical American experience.

Tragically, the cities where the urban poor live are often among the worst places in America to do business. Poor neighborhoods tend to be expensive places to operate, with heavy city taxes, lots of red tape from the city bureaucracy, few skilled employees, and a limited market. Months before the city of Detroit went bankrupt in 2013, the *Detroit Free Press* reported on Lisa VanOverbeke's attempt to open a cycling studio in the city. Her exercise business had already been successful in the suburbs, but "VanOverbeke didn't count on a series of last-moment requests, fees and municipal red tape that doubled the start-up costs when compared to her Royal Oak and Rochester locations." The obstacles proved too much: "VanOverbeke learned at the last moment that she would need to pay an extra $27,000 to soundproof the room to shield apartment dwellers from loud music during cycling classes. And that meant another city inspection and fee. Frustrated, she pulled out of the deal."[27] Detroit, in its spiral toward bankruptcy, lost the business.

To remove these barriers to job creation for the poor, Jack Kemp's proposal for no-tax free-enterprise zones should be resurrected for the poorest neighborhoods. Cities should be challenged to create red tape–free zones as the prerequisite to becoming a tax-free zone. We also need a major initiative in Washington to eliminate all but the most essential red tape for small businesses. And Obamacare has to be repealed because no jobs program will work as long as the law is putting such a burden on employment.

A new American breakout would help the poor more than any other group. And it can happen. It will require conservatives to care enough about their fellow citizens to engage in conversations that will often make both sides uncomfortable. It will require liberals to care enough about their fellow citizens to challenge some of their own most deeply held beliefs and some of their strongest allies.

Today, poverty threatens one out of five Americans, and that figure might rise to one out of four. That's an unacceptable future, and it ought to blast us out of our complacency and our assumptions.

CHAPTER ELEVEN

BREAKOUT IN ACHIEVING CURES

As we saw in chapter three, we are on the verge of an extraordinary breakout in health. Developments in science and technology could transform how long we live, how well we live, and how independently we live. For individual Americans, this could be a breakout of enormous importance.

The implications for the government are enormous as well, and the topic deserves more thought than our elected officials have given it. Medical costs account for such a big portion of the government's expenditures—between Medicare, Medicaid, and dozens of smaller programs and regulatory agencies, not to mention Obamacare—that curing the most common diseases will transform government.

If we can reduce to trivialities the real human problems for which many of our largest programs were established, we will find ourselves in a very different world. The problems of Medicare, Medicaid, health insurance coverage, and even Social Security—all of which appear daunting today—will assume an entirely different appearance, if they don't disappear entirely.

Consider just one common malady: Alzheimer's disease. A few years ago, I was privileged to serve on the Alzheimer's Study Group. We discovered that this disease alone will cost taxpayers $20 *trillion* from now till 2050. This cold number hardly conveys the human pain and exhaustion that the disease inflicts on patients, families, and caregivers. But you can also imagine how quickly the fiscal burden of millions of Alzheimer's patients adds up. Understanding Alzheimer's enough to postpone its onset by only five years could cut those costs in half because it is largely (though not entirely) a disease of the elderly.

You can probably come up with several other medical conditions—diabetes or heart disease, for instance—that will cost taxpayers trillions in coming decades.

Almost nothing we could do to save taxpayers' money would compare with the savings of curing the most common and expensive diseases. That's because cures can dramatically reduce the cost of healthcare and of sustaining people in the later years of their lives. Such cures would also ensure that America remains the most medically advanced country in the world, creating jobs and keeping the United States a hub of innovation. Achieving that breakout could be among the most-effective ways to control the federal budget over the long term.

The opportunity to cure so many diseases arises from the interaction of four separate but parallel revolutions that could reinforce and magnify each other in ways few of our political leaders seem to understand. The four revolutions are in genetics, regenerative medicine, advanced brain diagnostics, and big data for health analysis.

Let's consider each one separately.

First, the genetics revolution is providing new insights into how bodies (human and otherwise) work. We are still in the early stages of discovering and applying this knowledge, but over the next few decades, this field alone will yield tremendous breakthroughs in curing diseases. Even more exciting, it may allow us to anticipate diseases and "turn them off" before they start. So exciting is the potential that my friend Dr. Andy von Eschenbach, a former commissioner of the Food and Drug Administration, believes we could cure cancer within fifteen years.

The genetics revolution could also help us develop new levels of prenatal care to preempt certain genetically driven birth defects. In some ways the discovery that folic acid for the mother can help prevent spina bifida was an early forerunner of the kind of preventive prenatal care that could soon become much more sophisticated.

Second, regenerative medicine is still in the takeoff stage. But as the work of Dr. Anthony Atala demonstrates, within a decade we could begin to routinely replace whole organs by taking your own cells and growing a new one for you.

Third, as the example of Alzheimer's suggests, advanced brain diagnostics is probably the greatest opportunity we have to truly transform the quality of life for tens of millions of people. Our brain is the most complex system we know of. Breakthroughs in understanding the brain could produce great strides in curing autism, Alzheimer's, Parkinson's, mental health problems, and a wide range of other conditions. These mental conditions can distort our lives as decisively as any physical condition. Depression alone is a major contributor to total health costs, as depressed individuals are much more likely to manifest other illnesses.

And fourth, "big data," the ability to collect and analyze huge amounts of information, has enormous implications for health research. As we develop more and more electronic health records, we will have the ability to aggregate huge amounts of data and analyze them. Electronic epidemiology will become a rich source of health information.

These breakthroughs are real. The question is whether government will hinder them, as it is doing now, or accelerate them, which it could do even while spending less than it does today.

Taxpayers should demand six important steps from the government immediately.

One: FDA Reform

As we saw in chapter three, the FDA is a major prison guard stopping the breakout in health. It must be overhauled. Its standards for evaluating new treatments and therapies are completely unsuited for the age of regenerative medicine, and the cost of getting these treatments approved in the United States will be absurdly high even compared with the absurdly high cost of getting normal drugs approved. There is a grave danger that breakthroughs in regenerative medicine will be made in American laboratories and then be introduced to patients in China, India, Japan, and Europe because FDA approval is too time-consuming and too expensive.

Furthermore, FDA reform must address the agency's lethally low tolerance for risk. In addition to allowing Americans to obtain non-FDA-approved treatments with informed consent, the agency should begin to loosen its requirements in an age when we could track side effects and bad results in real time and make changes based on immediate data. Perhaps good preliminary results should be sufficient for doctors to begin prescribing many treatments as long as patients are connected to the constant monitoring technologies that Dr. Eric Topol described.

Two: Scoring Lifetime Savings instead of Annual Savings

As we begin to develop therapies that solve chronic conditions, government needs to develop new methods of scoring the costs of those therapies. Take, for example, the cost of kidney dialysis versus the cost of growing a new kidney. Kidney dialysis may be the cheaper treatment if you look at the budget for a single year, as Congress is apt to do. Once you are on dialysis, however, you remain on dialysis for the rest of your life. A debilitating experience, it requires two or three days a week at the clinic and the rest of the time spent recovering your strength. While the procedure may prolong your life, it also eliminates much of your ability to enjoy life and to earn a living. A regenerative kidney, on the other hand, would restore you to full health and enable you to return to work and to enjoy life to its fullest.

The challenge today is that a one-year score of the cost of dialysis versus a kidney replacement makes dialysis look less expensive. On the other hand, a lifetime cost of dialysis could soon make a regenerative kidney look like a bargain. This regenerative advantage becomes even more obvious if you include the difference between losing your job and depending on government payments and keeping your job and paying taxes.

There will be an explosion of research efforts if we can get to a fair scoring system for lifesaving breakthroughs.

Three: Prizes

Prizes should be developed for the most-important cures. As I discuss in depth in chapter seven, prizes can play a major role in raising funds, encouraging pioneers, and achieving big breakthroughs faster and cheaper than traditional bureaucratic efforts. *Taxpayers pay nothing* unless someone actually makes the defined breakthrough. At that point, we save money since we have technology that cures an expensive disease. An inexpensive kidney replacement to eliminate kidney dialysis would pay for itself in the first year or two. A cure for Alzheimer's, too, would pay for itself almost immediately. Surely these breakthroughs would be worth substantial prizes.

Four: Creating Special Bonds and Taking Key Research Off-Budget

We should seriously consider taking major areas for basic research off-budget and issuing special bonds to pay them off. Today, we fund vital research at the pace the overall budget and the overall competition for resources allow. Yet if these fields of research could save tens of billions (or in some cases trillions) of dollars, we should not treat them the same way we treat standard earmarks. We should try to reach them as fast as possible. Every year of delay is money lost.

Instead of standard budgeting, we could fund them with money raised from disease-designated bonds. When the breakthroughs occurred that enabled us to save money, a fraction of the savings could go to pay off the bonds. This approach would liberate crucial basic research from the prison of congressional pork-barrel spending and enable us to achieve lifesaving breakthroughs as rapidly as possible.

Five: Reorganize the National Institutes of Health

The National Institutes of Health need a thorough reorganization to focus on the great breakthrough areas that could dramatically improve health. Each of the twenty-one institutes jealously guards its budget and its prerogatives. These include institutes for research in dentistry, "environmental health sciences," drug abuse, alcohol abuse, and "minority health and health disparities," among others. Much of the research is focused on esoteric topics with little prospect for achieving a real breakthrough. In fact, far too much of the NIH is a comfortable bureaucracy going through the motions, and the director does not have the ability to shift resources and create dynamic working groups in truly promising areas. The institutes' work should focus almost exclusively on basic research in areas like brain science, genetics, and cancer.

Six: Congressional Hearings and Continuous Learning for Health Decision Makers

If you were to test the bureaucracy that pays for healthcare and the bureaucracy that regulates healthcare on the state of the arts in medicine, you would almost certainly find tremendous gaps in understanding. Similarly, if you examined the members of Congress and their staffs who have oversight of health issues, you would find very fragmented understanding of the potential breakthroughs.

There are more scientists in the world today than in all of prior history combined. These scientists get better computers and better laboratory equipment every year. As they connect to each other all over the world, new discoveries spread with remarkable speed. Continuous learning will be essential for anyone trying to make decisions on health policy or to manage health outcomes. The American people must insist that public servants stay abreast of these changes.

Congressional hearings on the potential breakout may be one of the right venues for this learning. In the early nineteenth century, congressional hearings and reports played a surprisingly large part in opening up the West to settlement. In the twenty-first century, they could educate our leaders and the country about the breakthroughs that will help America break out.

CHAPTER TWELVE

BREAKOUT FROM DISABILITIES TO CAPABILITIES

In 2004, Tammy Duckworth was copiloting an army Black Hawk helicopter in Iraq when it was shot down by insurgents. She lost both her legs and severely damaged one of her arms.

Today, Tammy Duckworth is a Democratic congresswoman from Illinois. She didn't go from a double amputation to Congress by focusing on what she *couldn't* do. She made it there by focusing on what she *could* do—which turned out to be quite a lot.

In a recent hearing of the House Oversight Committee, Representative Duckworth sat face-to-face with Braulio Castillo, whose status as a disabled veteran earned his company preferential treatment from the government—contracts worth up to $500 million.[1] What qualified Mr. Castillo as a "disabled" vet? He'd injured his foot in the 1980s playing football at the U.S. Military Academy prep school before later going on to play college football. His exchange with Congresswoman Duckworth made headlines across the country.

"Mr. Castillo, how are you?" she began. "Thank you—thank you for being here today."

"I am not well, but you're welcome," Castillo replied.

"All right, so, your foot hurt, your left foot?"

"Yes, ma'am."

"It hurts. Yeah, my feet hurt too," Representative Duckworth said. "In fact, the balls of my feet burn continuously, and I feel like there is a nail being hammered into my right heel right now, so I can understand pain and suffering and how service [connected injuries] can actually cause long-term, unremitting, unyielding, unstoppable pain. So I'm sorry that twisting your ankle in high school has now come back to hurt you in such a painful way, if also opportune for you to gain this status for your business as you were trying to compete for contracts. I also understand why, you know, something can take years to manifest [itself] from when you hurt [it]. In fact, I have a dear, dear friend who sprayed Agent Orange out of his Huey in Vietnam, who, it took forty years, forty years, for the leukemia to actually manifest itself, and he died six months later, so I can see how military service, while at the time you seem very healthy, could forty years later result in devastating injury. Can you tell me if you hurt your left foot again during your football career, subsequently to twisting it in high school?"

Castillo pretended not to understand the question.

"You played football in college, correct?" Representative Duckworth asked.

"Yes, ma'am."

"As a quarterback?"

"Yes, ma'am. I did."

"Did you hurt, did you injure that same foot again subsequently in the years since you twisted it in prep school?"

"Not to my recollection, ma'am."

Castillo proceeded to try to explain that despite playing football in college, he'd suffered pain from the original ankle injury for years.

"Do you feel the 30-percent rating that you have for the scars and the pain in your foot is accurate to the sacrifices that you've made for this nation?" Representative Duckworth asked. "That the VA decision is accurate in your case?"

"Yes ma'am, I do," he replied.

"You know, my right arm was essentially blown off and reattached," she said. "I spent a year in limb salvage with over a dozen surgeries over that time period, and in fact, we thought we would lose my arm, and I'm still in danger of possibly losing my arm. I can't feel it; I can't feel my three fingers. My disability rating for that arm is 20 percent."

Representative Duckworth then read from a letter Castillo had prepared. "My family and I have made considerable sacrifices for our country," he had written. "My service-connected disability status should serve as a testimony to that end. I can't play with my kids because I can't walk without pain. I take twice-daily pain medication so I can work a normal day's work. These are crosses—these are crosses—that I bear due to my service to our great country, and I would do it again to protect this great country."

"I'm so glad that you would be willing to play football in prep school again to protect this great country," Representative Duckworth concluded. "Shame on you, Mr. Castillo. Shame on you."[2]

That exchange illustrates the two ways people can approach disability. On one side was a citizen who viewed her severe injuries as a challenge to overcome but not an end to a productive and complete life. On the other side was a man who viewed the smallest physical difficulty as a qualification for taxpayer support.

In a time of increasing medical capability and an economy based less and less on physical labor, you might expect citizens to face the everyday pains with a fraction of Tammy Duckworth's courage and determination. Unfortunately, an increasing number of Americans are trapped in the disability prison, viewing physical challenges (serious or trivial) as a reason or an excuse to end their productive lives.

National Public Radio recently aired an alarming series of reports, *Unfit for Work*, documenting the rise of disability in America. The journalist Chana Joffe-Walt visited Hale County, Alabama, where *one in four* residents receives a disability payment from the Social Security

Administration each month. She spoke with many of Hale County's "disabled" residents and found some who did seem to have serious physical impairments. More commonly, though, they had the standard "back pain," or other manageable health problems like high blood pressure or diabetes. Labeling them as "disabled," it seemed, stretched the meaning of the word. Yet as Joffe-Walt reported, "Over and over again, I'd listen to someone's story of how back pain meant they could no longer work, or how a shoulder injury had put them out of a job. Then I would ask: What about a job where you don't have to lift things, or a job where you don't have to use your shoulder, or a job where you can sit down? They would look at me as if I were asking, 'How come you didn't consider becoming an astronaut?'"[3]

Millions of Americans are now receiving disability payments from the federal government—fourteen million, to be precise. For comparison, that's more than the population of the thirteen smallest states put together; only four states in the country have more than fourteen million people. Cash payments to these "disabled" Americans now cost taxpayers more than food stamps and welfare combined.[4]

The overwhelming majority of disability recipients do not work. An analysis of federal survey data by the *Washington Examiner* found that only 13 percent have worked for pay since they started receiving benefits.[5] And once they enroll in the program, they're unlikely ever to return to independence.

"People who leave the workforce and go on disability qualify for Medicare," Joffe-Walt reports. "They also get disability payments from the government of about $13,000 a year. This isn't great. But if your alternative is a minimum wage job that will pay you at most $15,000 a year, and probably does not include health insurance, disability may be a better option."

Welfare reform had the unanticipated consequence of giving states the incentive to move people from welfare (where the states pay part of the cost) to disability (where the federal government picks up the cost). Clever entrepreneurs have even established companies to help states move people from their welfare rolls to the federal government's disability rolls.

Most disability recipients aren't as corrupt as Tammy Duckworth's sparring partner. Many are just trying to survive in a bad economy.

In Joffe-Walt's conversations with "disabled" Hale County residents, one name came up over and over again: Perry Timberlake, a local physician. When she visited his office, she discovered that though he wasn't exactly running a disability scam, he was basing his subjective decisions about which patients qualify as "disabled" on some questionable criteria.

"We talk about the pain and what it's like," he told her. "I always ask them, 'What grade did you finish?'" Of course, this question has nothing to do with whether a patient has a debilitating health problem. But the doctor told Joffe-Walt that he considers the information crucial in evaluating the patient's ability work in rural Hale County. "Dr. Timberlake," she says, "is making a judgment call that if you have a particular back problem and a college degree, you're not disabled. Without the degree, you are."[6]

The doctor's comments are striking for two reasons. First, they confirm the worst stereotypes of some disability recipients—on the dole even though they aren't truly disabled. But second, while the doctor shouldn't be using the program to hand out welfare, he has a point. It has become difficult to find work in many places if you don't have a high school education. Disability handouts for the non-disabled are not the solution to this problem, however. They only obscure it.

The disability program isn't just trapping these Americans in permanent poverty; it's also trapping a new generation with the same educational disadvantage. Families have learned that if their children are diagnosed with mental difficulties that affect their performance in school, they can collect disability for their kids as well. For a poor family, that seven hundred dollars per month per child may be its main source of income. Many families' incomes, therefore, are perversely tied to their children's falling behind in school.

The problem isn't small. Thanks to these incentives, the incidence of children with disabilities has increased by 700 percent in the last thirty years.[7] Crippling a child's ability to live a full life for the sake of a seven-hundred-dollar monthly check is child abuse of almost Dickensian proportions.

Eight former Social Security commissioners objected to the *Unfit for Work* series. They argued that most of the increase in disability payments was due to demographic changes, an aging population, and the greater involvement of women in the workforce.

The commissioners' response was a classic example of the bureaucratic mindset. They did not explain why, after two generations of progress in treating veterans with disabilities, there has been no corresponding progress for civilians. And they ignored the incentives for individuals, state governments, and businesses to abuse the federal disability programs.

The *Examiner*'s recent analysis of federal survey data further undermines the commissioners' objections. As the *Examiner* summarized in its findings, "Recipients of federal disability checks often admit that they are capable of working but cannot or will not find a job, that those closest to them tell them they should be working, and that working to get off the disability rolls is not among their goals. More baffling, most have never received significant medical treatment and [have] not seen a doctor about their condition in the last year, even though medical problems are the official reason they don't work." The data showed that returning to work is not a goal for over 70 percent of people on Social Security Disability Insurance. At the time the survey was taken, 96 percent had not looked for work in the previous four weeks. Only 25 percent saw themselves working in the next five years.[8]

This corrosive system of dependency contradicts the foundational American principle that we are endowed by our Creator with the unalienable right to the pursuit of happiness. It is obscene that the government erected to secure that right encourages mothers to sabotage their children's education.

The corruption of federal disability programs is a tragedy for all involved, especially those who are cheated out of the dignity and fulfillment of a productive life. The government of the United States should not be in the business of narrowing its citizens' horizons. This corruption is also a financial disaster. Bestowing a lifetime of subsidies on someone because of a one-time diagnosis of disability makes no sense.

The remedy is not to "get tough" with disability cheats. Tighter regulation, more aggressive reviews, and longer waiting periods will not solve the problems that are baked into the system. We must change our focus from what limits people to what empowers them.

With the amazing breakthroughs in health and in lifelong learning, we have the opportunity to change the obsolete concept of disability into the new concept of capability. We have to rethink the system.

A system of lifelong learning—built on innovations like Khan Academy, Kaplan, and Udacity—should allow every American to learn what he needs to have a decent job. If you can't keep your warehouse job because of a back injury, there are hundreds of other trades or careers you can learn instead. Our policy should reflect and reinforce these opportunities. Dr. Timberlake's automatic disability certifications should become a thing of the past, since anyone who doesn't have an extremely serious health problem should be able learn the skills he needs to get a job he can do.

Breakthroughs in health should make many forms of disability obsolete, too. If you require kidney dialysis, there's a good chance you qualify for disability payments; your medical condition really does make it difficult to work. But if the work of pioneers like Dr. Anthony Atala succeeds in regenerating kidneys, that disability will go away.

Helping people with serious challenges get to and from work may be one of the first areas in which we deploy self-driving cars. We already build advanced systems to help seriously disabled veterans. This technology should soon be available to all Americans.

We could turn fourteen million people from passive recipients of government aid into self-sufficient, income-earning citizens, people who exercise the American birthright of the pursuit of happiness.

CHAPTER THIRTEEN

BREAKOUT CHAMPIONS

Americans have a reason for hope. We have the opportunity to enter a new era in which we live longer, healthier, more independent lives; in which easily affordable lifelong learning gives us unprecedented security and flexibility in our careers; in which prosperity and opportunity expand as never before.

But it will take a lot of work to make it happen.

The pioneers of the future have already achieved many breakthroughs, but there are large areas of American life in which the prison guards of the past are blocking breakout.

We need a popular movement to overwhelm the prison guards. Moving our laws and government away from the litigation-focused, bureaucratic model of the past will require *breakout champions*—the citizens, elected officials, candidates, and government staff who push for the changes we need to achieve breakout.

How Breakout Champions Think about Government and Public Policy

Our society, in this era of big government, is centered on bureaucracy, regulation, and law. This mindset makes breakthroughs difficult and a general breakout almost impossible.

To become breakout champions, citizens and elected officials need to think about society and government in a completely new way. There are eight key principles that distinguish pioneers of the future and the breakouts they bring about.

One: Look beyond Government

Leaders in this new breakout world, in which new technologies empower ordinary citizens, realize that there are many things that can be solved by free people banding together in creative ways. Bureaucratic government should be the last resort. When looking for solutions or measuring resources, start with society, not the bureaucracy. When organizing solutions, start first with voluntary, decentralized organizations, and go to bureaucracies and government only as a last resort.

Two: Focus on Outcomes, Not Input

Because a breakout requires dramatically smaller bureaucracies and fewer centralized regulations, our systems should focus much more on outcomes than inputs. For example, how many children have learned to read is a more important question than how much is being spent on the reading program. In a decentralized society, customers are important, and customers care about outputs, not inputs. The result is constant pressure for better quality and lower cost. Since a decentralized system is open to many solutions, the ideas that work can bubble up while the ideas that fail can shrivel up. Big bureaucracies do the opposite: they escape accountability and simply demand more resources, so failure becomes an excuse to build an even bigger system with even more expensive failures.

Three: Continually Work on Improvements

Breakouts emerge from pioneers working independently or in small groups on breakthroughs. They constantly drive for improvements. The bureaucratic system of cumbersome, slow, and rigid central planning guarantees that the institutions that would be improved by the breakthroughs will not adopt them. And as bureaucracies become less competent, they defend themselves by excluding, opposing, or even destroying what the pioneers are offering. The answer is not to try to build the perfect breakthrough replacement. The answer is to design systems capable of continuous improvement and open to constant internal challenges and adaptations—systems, that is to say, that are the opposite of the bureaucracies we have inherited from the recent past.

Four: Look for Solutions in Unusual Areas

Most so-called experts are people who know more and more about narrower and narrower topics. In an age of breakout, however, new developments in one area can be applied to problems in another area. Winston Churchill, while First Lord of the Admiralty, invented the tank after realizing that the combination of the internal combustion engine and treads could be a solution to the problems of trench warfare. No one in the *army* had thought of this. The question of using smartphones and apps in medicine or education or public safety is a more recent example. Dramatic change occurs when people working in one specialty are alert to what they can learn from advances in other fields.

Five: Imitation Is Almost Always Cheaper than Invention

People in positions of responsibility in both the private and public sectors should spend at least one day a month exploring the new breakthroughs in fields other than their own, asking themselves if those advances could work for them. It will almost always be less expensive in time and money to import solutions than to invent them yourself.

Six: Remember to Look for Opportunities as well as Solutions to Problems

Breakthroughs often create opportunities that have nothing to do with problems they were intended to solve. For example, the smartphone opens up opportunities in education, medicine, public safety, commerce, and work. The fax was not just a better way to send a letter. The internet is not just a better way to send a fax. Each of these advances created opportunities that no one had thought of until the technology actually existed. We may achieve more progress out of opportunities-focused analysis of new breakthroughs than if we devoted the same amount of time narrowly searching for solutions to currently understood problems. We need to do both, but we usually spend far more time and energy on problems we already understand than on new opportunities.

Seven: Individual Incentives Matter

Because every American is a potential problem-solver, a lot of thought must go into aligning values and incentives. In changing our focus from the centralized bureaucracy to the free society that fosters the pioneers of the future, it is important to appreciate the incentives we're establishing. Microeconomics (how individuals, small businesses, and markets behave) becomes more important than macroeconomics (how economies as a whole behave) in thinking through how to lead a healthy, prosperous, self-reliant, and free society.

Eight: Study Previous Breakouts

We can gain valuable insights by studying the previous periods of tremendous scientific, technological, and entrepreneurial achievement in American history. How did public officials and leading citizens navigate those momentous technological journeys? What mistakes did they make that we can avoid?

How Public Officials Can Be Breakout Champions

At every level of government, from the local school board to the U.S. Congress, public officials can take some important steps to encourage the next American breakout.

A public official, first of all, should be a teacher for the community. The words a legislator uses, the issues she focuses on, and the meetings she puts into her crowded schedule are powerful signals for others. Simply talking about the breakthroughs of the pioneers of the future and how prison guards of the past use government to halt progress encourages new ways of thinking. In their communication with their constituents, legislators should draw attention to the breakthroughs of pioneers in their district, across the United States, and around the world.

In particular, every legislator should develop an advisory committee of persons under the age of eighteen and should have at least one meeting per quarter in which any young person with ideas for the future or examples of pioneering activity could come and talk. Asking the young what they think can be done to develop new solutions and use new technology will break us far outside the restricted thinking and ingrained habits and institutions of the old order.

The town hall meetings that have become a popular means of communication with constituents are a good place to highlight the breakthroughs of the pioneers of the future and to shine a spotlight on the prison guards who would thwart them. A focus on breakthroughs gives a positive, solutions-oriented tone to the town hall meeting, in contrast with the hostility and conflict that sometimes characterize these encounters with the voters.

How Private Citizens Can Be Breakout Champions

You do not need to hold office to be a breakout champion. Every citizen can contribute.

In this age of social media, when Americans get so much of their news from family and friends, a simple first step to becoming a citizen breakout champion is sharing information on Facebook, Twitter, and other platforms. When sharing the story, make it vivid by telling how that breakthrough might make life in your community better. You will develop a circle of friends and associates who are passionate about breakout.

If you belong to a civic club or meeting group, invite pioneers of the future to speak. Educate your elected representatives about breakout and demand that they enact the changes we need.

When citizens go to a town hall meeting and ask questions about pioneering developments, they are accelerating the rate of change.

When citizens write letters to the editor identifying prison guards of the past and the cost of their opposition to a better future, they are accelerating the rate of progress.

When citizens write or call their elected officials, they are drawing their attention to new ideas and new developments or to roadblocks and limitations. Public awareness often precedes and leads official awareness.

When citizens draw attention to the breakdown of the big bureaucratic system and demand transparency to identify the waste and fraud in government, they are moving us toward a new breakout model of effective, affordable self-government.

Finally, when you are asked to support candidates or are looking for candidates to support, find out if they understand the principles of breakout. Do they appreciate pioneers of the future? Are they committed to breakout policies in education, energy, and health?

I will be offering workshops, short courses, and online working groups for citizens who want to get more deeply engaged. Go to www.BreakoutUniversity.com, and you can be engaged with others in identifying and helping pioneers of the future and identifying and defeating the prison guards of the past.

CONCLUSION

The past few years have been difficult for many Americans.

As the financial crisis of 2008 gave way to a prolonged economic slowdown, shock turned to fear and finally to hopelessness. Americans have wondered if higher unemployment, lower incomes, and diminished opportunity are here to stay. Some have accepted these changes as the "new normal."

I hope this book has convinced you that an America with millions of people looking for work, failing schools, rationed and bureaucratic healthcare, dwindling and costly sources of energy, and broken-down government is not our inescapable future. There is a way out.

The future could be amazing. The innovations we are seeing in learning, health, energy, transportation, and even government have the potential to transform our lives. They promise to make America freer and more prosperous, with real opportunity for all. They promise to make life better and richer. They promise, in short, to achieve the American dream.

This is the choice America faces: the choice between breakout and breakdown.

Will we accept schools that do not educate our children and colleges that burden them with debt, or will we embrace the better alternatives?

Will we endure an inadequate healthcare system that gets more expensive every year, or will we speed the breakthroughs that can produce revolutionary cures and bring down costs?

Will we develop our country's vast energy resources, or will we continue to pay unnecessarily high prices at great human cost?

Finally, will we accept an overbearing government, or will we renew the authority of the people over the bureaucratic state? This is really the fundamental question. The choice between breakout and breakdown is a political choice.

We have seen how the prison guards of the past use centralized bureaucracy, litigation, regulations, and red tape to delay or kill breakthrough innovations in many fields. They squander America's potential in order to protect their privileges and their old ideas, and they rely on our complacency not to do anything about it.

In his first inaugural address, Ronald Reagan said, "It is time for us to realize that we are too great a nation to limit ourselves to small dreams. We're not, as some would have us believe, doomed to an inevitable decline. I do not believe in a fate that will fall on us no matter what we do. I do believe in a fate that will fall on us if we do nothing." Our country faces much the same decision today.

America needs a breakout movement, a new politics that is less about the Right versus the Left than about the future versus the past. It will require a new political language, new coalitions of pioneers and champions who understand the importance of breaking out, and new assertiveness in exposing the prison guards of the past.

The bureaucracies and systems of the old order cannot be reformed. They must be replaced. But if, in our effort to dismantle them, we focus only on what is wrong, we will never gain the momentum and public support needed for real change.

Fortunately, the very breakthroughs the prison guards are fighting to stop are the model for their replacement. Learning science and individualized education, regenerative and personalized medicine, and abundant energy and self-driving cars, among many other groundbreaking innovations, could solve so many of the problems Americans face today. I hope we will see them in my lifetime and yours. That will depend on your help.

ACKNOWLEDGMENTS

Work on *Breakout* took several months, but the project really goes all the way back to my presidential campaign and even to my days in Congress. As a candidate for president, I was struck by the difficulty of getting the political press to cover a topic as consequential as brain research or finding cures for Alzheimer's. Although these issues mattered to millions of Americans, the press corps would not cover them, no matter how positive a response we got on the ground.

Following that campaign, I noticed that this response was part of a much larger pattern, and I was increasingly struck by the disconnect between the amazing innovations occurring outside of Washington—changes that seemed to have large implications for public policy—and the city's preoccupation with the day-to-day trivia of politics.

At Gingrich Productions we were so impressed with the difference between Washington's assumptions and a rapidly changing America that we undertook an extensive review of why Republicans were so surprised by the 2012 election.

The more we looked at the gap between the assumptions and the reality, the more we realized that something profound was happening among the American people. Realities outside Washington simply weren't penetrating the Washington elites in either party, in the bureaucracy, or in the political news media.

What truly amazed us was the gap created by Americans' creativity, excitement, enthusiasm, and problem solving. The country was pulling away from Washington, and the gap was getting bigger every year.

The executive summary of lessons learned from 2012 can be read at www.GingrichProductions.com/lessonslearned.

I want to acknowledge the support of Stephen Bonner, August Busch III, Frank Hanna Jr., Rick Jackson, Terry Kohler, Mario Kranjac, Mike Leven, Mike Murray, Christopher Ruddy, George and Breda Shelton, and Harold Simmons, without whom the months of research, exploration, and thought would have been impossible. Their help was the necessary bridge to conceptualizing what a breakout could look like.

The first sign of this breakout was the fracking revolution. I spent many hours discussing the revolution in American energy with my friend Scott Noble, the founder and CEO of Noble Royalties. While many in Washington continued to emphasize expensive alternative energy programs, Scott understood how deep and powerful the revolution in oil and gas production would be. Scott's expertise and advice on this project has been extraordinarily helpful.

By chance I met separately with Sebastian Thrun and Bror Saxberg, each of whom told me about the amazing changes coming to education, which had scarcely entered the political discussion.

Sebastian also told me about his work at Stanford and Google on self-driving cars, which impressed me even further as I began to think about the transformative potential. Later, Robert Norton visited our office and made quite an impression with his stories of the barriers to automotive innovation.

I learned some years ago about Dr. Anthony Atala's groundbreaking work in regenerative medicine and benefited enormously from his briefings on the latest science.

Finally, Alex Castellanos recommended Lieutenant Governor Gavin Newsom's book *Citizenville*, which provided a compelling glimpse of the extraordinary opportunities for innovation in government.

Somewhere in the process of discovering all of this potential and contemplating the barriers to embracing it, we realized we had to explain the great opportunity in a book so people could see what was possible.

We are now hard at work developing online learning at www.BreakoutUniversity.com to continue collaborative learning with every citizen who wants breakout to become reality.

Ross Worthington has been an invaluable partner in developing *Breakout*. This book reflects his intelligence, talent, and many hours of hard work. Among other things Ross edits my e-newsletter, from which a few sections of this book have been adapted and which you can sign up to receive at www.GingrichProductions.com.

Ross and I were helped tremendously in this project by very smart and dedicated people.

Vince Haley and Joe DeSantis spent countless hours with us thinking through the *Breakout* concept at the whiteboard and developing the argument of the book. Joe made especially significant contributions to the energy chapter, and it was Vince who originally summarized our goal as a breakout from the emerging "new normal." Everything became simpler after that. The project would have been impossible without them.

The support of the rest of our hardworking colleagues at Gingrich Productions was equally essential. Bess Kelly is at the center of everything that Callista and I do and is indispensable. Christina Maruna ably leads our online and social media efforts, while Woody Hales, John Hines, and Taylor Swindle round out a truly exceptional team.

Interns Allison Coyle, Jamie Greedan, Jack O'Donnell, and Angela Vargas (along with the aforementioned John Hines) contributed substantial research.

Terry Balderson, a research advisor who for years has sent me hundreds of news articles from around the world each day, contributed many pieces of information. Indeed, his support allows me to stay up to date on a constantly changing world. Jeremy Vaught, our all-around web guru, also contributed a number of good ideas and examples reflected in this book.

Tom Spence and Katharine Mancuso at Regnery did a terrific job editing and greatly improved the book. They put in many hours of hard work to meet a tight deadline. Marji Ross, president and publisher at Regnery, and Jeff Carneal, president of Eagle Publishing, both provided tremendous support for a project that was completely different from any other we've worked on together over the years. As always, I am grateful for their help.

Craig Shirley has tied a lot of my thinking firmly into the long evolution of conservatism starting with President Reagan, Congressman Jack Kemp, and myself in the 1970s. As Craig reminds us, this is in many ways the culmination of forty years of working to get beyond the failed policies and ideas that have been dominant. Peter Ferrara has been an amazingly creative thinker throughout this period and challenged us again and again to be bold.

Republican National Committee chairman Reince Priebus was very supportive in hosting Newt University at the 2012 Republican National Convention and then in including the Gingrich Productions team in his own "lessons learned" effort, later inviting me to discuss some of the ideas in *Breakout* at the RNC's summer meeting. He has shown real commitment to figuring out how the party can "break out." His chief of staff, Mike Shields, has facilitated much of our effort.

Congresswoman Cathy McMorris Rodgers, too, reinforced the *Breakout* concept with great enthusiasm and displays remarkable vision and innovation as chair of the House Republican Conference. I also want to thank Congressmen Mike Burgess, Tim Griffin, Trent Franks, Tom Price, Greg Walden, and Pete Sessions for their encouragement. I want to thank Senator Marco Rubio for his future-oriented leadership as Speaker of the Florida House. In addition, Senators Rob Portman, Ron Johnson, Rand Paul, Mike Lee, and Ted Cruz have all encouraged the development of bold solutions. I also want to thank Governors Nathan Deal, Rick Perry, John Kasich, Bobby Jindal, Bob McDonnell, and Scott Walker. Their encouragement and their example of innovative leadership and effectiveness kept me moving forward.

My old friend from my days in the House, former Congressman Bob Walker, is just as future-oriented and excited about science and technology as I am, and his advice on this project was very helpful, especially on the space chapter (a particular passion of his).

My great friend Tony Dolan, who served as President Reagan's chief speechwriter for eight years, contributed much valuable feedback and advice.

As always, Randy Evans and Stefan Passantino assisted us with their legal expertise. Randy has long been my general advisor on everything, and Stefan is an excellent counsel.

I am deeply grateful to my two daughters, Kathy Lubbers and Jackie Cushman. As president and CEO of Celebrity Leaders, Kathy does a terrific job of representing me. She has guided Callista and me through a combined twenty-seven books (including fifteen *New York Times* bestsellers) and seven documentary films. Jackie is a gifted writer, the author of two books, and an insightful columnist for *Townhall*. She is also a wonderful mother to our two grandchildren, Maggie and Robert. Kathy and Jackie's husbands, Paul Lubbers and Jimmy Cushman, are always supportive of our next adventure.

Finally, nothing I do would be possible without the love and support of my wife, Callista. Callista is a successful author herself, the creator of the *New York Times* bestselling children's series featuring Ellis the Elephant, teaching four-to-eight-year-olds American history. In her latest book, *Yankee Doodle Dandy*, Ellis discovers the American Revolution. In addition to her writing and her leadership as president of Gingrich Productions, Callista is a movie producer, cohost, narrator, and photographer. She also devotes hours each week to singing at the Basilica of the National Shrine of the Immaculate Conception and to playing the French horn in the City of Fairfax Band. I am enormously proud of her accomplishments and grateful for her support in everything we do.

NOTES

Preface

1. James F. McCarty, "Ex-VA Director William Montague Expected to Plead Guilty to Bribery, Fraud, Conspiracy Charges," *Plain Dealer*, available online at Cleveland.com, February 20, 2014, http://www.cleveland.com/countyincrisis/index.ssf/2014/02/ex-va_director_william_montagu.html.

Introduction

1. Julie Hoffer, "Battling for Life," *Cavalier Daily*, April 27, 2001, http://www.cavalierdaily.com/article/2001/04/battling-for-life/.
2. Ibid.
3. Steven Ginsberg, "One Life Galvanizes Thousands: Out of Options, Virginia Woman Fights for Experimental Cancer Drugs," *Washington Post*, May 7, 2001.

4. "Timelines: 1983; C225," Life Sciences Foundation, accessed August 2013, http://www.lifesciencesfoundation.org/events-C225.html.
5. Reuters, "Rise and Fall of ImClone Systems," Fox News, June 12, 2002, http://www.foxnews.com/story/0,2933,55120,00.html.

Chapter One

1. Daniel Walker Howe, *What Hath God Wrought: The Transformation in America, 1815–48* (New York: Oxford University, 2007), 526.
2. Andrew Martonik, "Larry Page: 1.5 Million Android Devices Activated Every Day," Android Central, July 18, 2013, http://www.androidcentral.com/larry-page-15-million-android-devices-activated-every-day.
3. "Tips and Statistics," Safety Insurance Group, accessed August 21, 2013, https://www.safetyinsurance.com/driversafety/tips_statistics.html.
4. Gavin Newsom and Lisa Dickey, *Citizenville: How to Take the Town Square Digital and Reinvent Government* (New York: Penguin, 2013).
5. The nineteenth-century French economist Frédéric Bastiat produced just such a satire, "A Petition From the Manufacturers of Candles, Tapers, Lanterns, Sticks, Street Lamps, Snuffers, and Extinguishers, and from Producers of Tallow, Oil, Resin, Alcohol, and Generally of Everything Connected with Lighting." His tongue-in-cheek letter was written well before the invention of the light bulb; he was asking the French government to ban the sun.

Chapter Two

1. Sarah Greek, "Bachelor's Degree—At Age 18," College Fix, July 11, 2013, http://www.thecollegefix.com/post/13981/?utm_source=feedburner&utm_medium=feed&utm_campaign=Feed.

2. "From Harshal," Khan Academy Stories, April 12, 2013, https://www.khanacademy.org/stories/harshal-april-12-2013.
3. Kara Miller and Kinne Cahpin, "Sal Khan Reinvents Education," WGBH News, April 26, 2013, http://www.wgbhnews.org/post/sal-khan-reinvents-education.
4. "From Dave," Khan Academy Stories, December 13, 2012, https://www.khanacademy.org/stories/dave-m-december-13-2012.
5. "MTT2K," YouTube video, posted by John Golden, June 18, 2012, http://www.youtube.com/watch?feature=player_embedded&v=hC0MV843_Ng.
6. Angela Chen, "Parody Critiques Popular Khan Academy Videos," *Chronicle of Higher Education*, June 28, 2012, http://chronicle.com/blogs/wiredcampus/parody-critiques-popular-khan-academy-videos/37543; Maggie Severns, "Math Teachers' Parody Prompts Project to Critique Khan Academy's Lessons," Slate, June 26, 2012, http://www.slate.com/blogs/future_tense/2012/06/26/khan_academy_mystery_science_theater_300_parody_by_math_teachers_video_.html.
7. Justin Reich, "Don't Use Khan Academy without Watching This First," EdTech Researcher (blog), *Education Week*, June 21, 2012, http://blogs.edweek.org/edweek/edtechresearcher/2012/06/dont_use_khan_academy_without_watching_mmt2k_first.html?intc=mvs.
8. Valerie Strauss, "Khan Academy: The Hype and the Reality," The Answer Sheet (blog), *Washington Post*, July 23, 2012, http://www.washingtonpost.com/blogs/answer-sheet/post/khan-academy-the-hype-and-the-reality/2012/07/22/gJQAuw4J3W_blog.html.
9. Strauss, "Sal Khan Responds to a Critic—and the Critic Answers Back," guest post by Karim Kai Ani, The Answer Sheet (blog), *Washington Post*, August 2, 2012, http://www.washingtonpost.com/blogs/answer-sheet/post/sal-khan-responds-to-critic/2012/07/25/gJQA83rW9W_blog.html.
10. Some estimates are even higher—the Evergreen Education Group estimates that 275,000 students in forty states are enrolled full-time in online classes. Vanessa Romo, "Are Charter Schools Really a

Good Idea?" Take Part, June 3, 2013, http://www.takepart.com/article/2013/06/03/virtual-school-charter-growth-America.

11. "High School Graduation Rate (1990–2012)," America's Health Rankings, 2012 Annual Report, http://www.americashealthrankings.org/all/graduation.
12. Yasmin Rammohan, "High Dropout Rate for Chicago Schools," WTTW.com, November 7, 2011, http://chicagotonight.wttw.com/2011/11/07/high-dropout-rate-chicago-schools.
13. "Reading Performance," National Center for Education Statistics, 2011, http://nces.ed.gov/programs/coe/pdf/coe_cnb.pdf.
14. "Leadership," Kaplan Inc., accessed August 2013, http://www.kaplan.com/about-kaplan/leadership.
15. Mechanical Turk allows developers to pay real people small amounts of money to complete simple tasks. For instance, you might pay someone three cents every time he confirms that the book cover shown on a product page indeed matches the title in the database. Amazon built the system for its own use and decided to open it up to anyone.
16. Michelle Jamrisko and Ilan Kolet, "Cost of College Degree in U.S. Soars 12 Fold: Chart of the Day," Bloomberg.com, August 15, 2012, http://www.bloomberg.com/news/2012-08-15/cost-of-college-degree-in-u-s-soars-12-fold-chart-of-the-day.html.
17. Jillian Berman, "Student Debt Balance around 60 Percent of Graduates' Annual Income on Average: Study," Huffington Post, June 19, 2013, http://www.huffingtonpost.com/2013/06/19/student-debt-graduates-income_n_3464924.html.
18. Andy Kessler, "Sebastian Thrun: What's Next for Silicon Valley?," *Wall Street Journal*, June 15, 2012, http://online.wsj.com/article/SB10001424052702303807404577434891291657730.html.
19. Ibid.
20. Steve Kolowich, "San Jose U. Puts MOOC Project with Udacity on Hold," *Chronicle of Higher Education*, July 19, 2013, http://chronicle.com/article/San-Jose-State-U-Puts-MOOC/140459/.

Chapter Three

1. "Aflac Real Cost Calculator Source Information," Aflac, updated January 2013, accessed August 2013, http://www.aflac.com/individuals/realcost/source/.
2. Ketaki Gokhale, "Heart Insurance in India for $1,583 Costs $106,385 in U.S.," Bloomberg, July 28, 2013, http://www.bloomberg.com/news/2013-07-28/heart-surgery-in-india-for-1-583-costs-106-385-in-u-s-.html.
3. Ibid.
4. Eric Topol, *The Creative Destruction of Medicine: How the Digital Revolution Will Create Better Health Care* (New York: Basic Books, 2012), vi. Italics added.
5. Ibid.
6. Ibid.
7. "iDoctor: Could a Smartphone Be the Future of Medicine?," *Rock Center with Brian Williams*, NBC News, January 24, 2013, http://www.nbcnews.com/video/rock-center/50582822.
8. Topol, *The Creative Destruction of Medicine*, ix–x.
9. Ibid., 20.
10. Ibid., 22.
11. Philip Klein, "The Empress of Obamacare," *American Spectator*, June 2010, http://spectator.org/archives/2010/06/04/the-empress-of-obamacare.
12. "Obamacare Rate Shocker: Committee Surveys Leading Insurance Companies—Obamacare to Cause Premiums to Spike Nationwide, as High as 400 Percent," United States House of Representatives Energy and Commerce Committee, May 13, 2013, http://energycommerce.house.gov/press-release/obamacare-rate-shocker-committee-surveys-leading-insurance-companies%E2%80%94obamacare-cause.
13. "Effects of the Affordable Care Act on Health Insurance Coverage," Congressional Budget Office, May 2012, http://www.cbo.gov/sites/default/files/cbofiles/attachments/44190_EffectsAffordableCareActHealthInsuranceCoverage_2.pdf.

14. Sharon Begley, "Delay in Obamacare Requirement Puts Onus on the Honor System," Reuters, July 5, 2013, http://news.yahoo.com/delay-obamacare-requirement-puts-onus-honor-system-053417869.html;_ylt=A2KJ2PZzwtZR.U0ABhPQtDMD%20%20target=.
15. Erin Gilmer, "Developing Mobile Apps a Medical Devices: Understanding U.S. Government Regulations," IBM.com, April 16, 2013, http://www.ibm.com/developerworks/mobile/library/mo-fda-med-devices/index.html.
16. Sarah underwent two double-lung transplants, as her first set of donor lungs sent her into primary graft failure, a complication that occurs in 10 to 25 percent of lung transplants. Doctors were able to successfully transplant a second set of "high-risk" lungs before Sarah's condition declined any further. See "Sarah Murnaghan Received Second Lung Transplant, Parents Say," Philly.com, June 30, 2013, http://articles.philly.com/2013-06-30/news/40273037_1_second-transplant-lung-transplant-graft-failure; Devin Kelly, "Pennsylvania Girl Got Second Lung Transplant After First Failed," *Los Angeles Times*, June 28, 2013, http://articles.latimes.com/2013/jun/28/nation/la-na-nn-sarah-murnaghan-second-transplant-20130628.
17. "Anthony Atala: Printing a Human Kidney," TED Talks, TED.com, March 2011, http://www.ted.com/talks/anthony_atala_printing_a_human_kidney.html.
18. Kevin Sack, "Kidney Transplant Committee Proposes Changes Aimed at Better Use of Donated Organs," *New York Times*, September 21, 2012, http://www.nytimes.com/2012/09/22/us/proposals-aim-to-improve-kidney-transplant-system.html?_r=0.
19. "Dialysis Overview," Glickman Urological and Kidney Institute, Cleveland Clinic, accessed August 2013, http://my.clevelandclinic.org/services/dialysis/np_overview.aspx.
20. "Diabetes Statistics," American Diabetes Association, updated August 20, 2013, http://www.diabetes.org/diabetes-basics/diabetes-statistics/.

21. "The Cost of Diabetes," American Diabetes Association, updated March 2013, http://www.diabetes.org/advocate/resources/cost-of-diabetes.html.
22. "What Organs Can Be Donated?," New York Organ Donor Network, accessed August 2013, http://www.donatelifeny.org/about-donation/what-can-be-donated/pancreas-transplants/.
23. "DATA: United States," New York Organ Donor Network, updated April 12, 2013, http://www.donatelifeny.org/about-donation/data/#Data%20US1.
24. Daniel Henninger, "Drug Lag," *The Concise Encyclopedia of Economics*, Library of Economics and Liberty, Liberty Fund, accessed August 2013, http://www.econlib.org/library/Enc1/DrugLag.html.
25. Ibid.
26. Avik S. A. Roy, "Stifling New Cures: The True Cost of Lengthy Clinical Drug Trials," *Project FDA*, report no. 5, Manhattan Institute, April 2012, http://www.manhattan-institute.org/html/fda_05.htm.
27. Ibid.
28. Ronald Trowbridge and Steven Walker, "The FDA's Deadly Track Record," *Wall Street Journal*, August 14, 2007.
29. "Diamond Food Inc. 2/22/10: Warning Letter," U.S. Food and Drug Administration, letter sent February 22, 2010, http://www.fda.gov/iceci/enforcementactions/warningletters/ucm202825.htm.

Chapter Four

1. Ted Gregory, "North Dakota's Oil Rush Lures Chicago-Era Residents," *Chicago Tribune*, March 17, 2013, http://articles.chicagotribune.com/2013-03-17/news/ct-met-north-dakota-fracking-chicagoans-20130317_1_oil-boom-oil-rush-man-camps.
2. Scott Tong, "George Mitchell, 94, Dies: Oil Man Unlocked Fracking," Marketplace, December 7, 2012, http://www.marketplace.org/topics/sustainability/oil-man-who-figured-out-fracking.

3. Tong, "George Mitchell, 94, Dies"; Michael Shellenberger, "Interview with Dan Steward, Former Mitchell Energy Vice President," The Breakthrough Institute, December 12, 2011, http://thebreakthrough.org/archive/interview_with_dan_steward_for.
4. Tong, "George Mitchell, 94, Dies."
5. Shellenberger, "Interview with Dan Steward, Former Mitchell Energy Vice President."
6. Institute for Energy Research, *North American Energy Inventory*, December 2011, http://www.energyforamerica.org/wp-content/uploads/2012/06/Energy-InventoryFINAL.pdf.
7. Jonathan Bailey and Henry Lee, "North American Oil and Gas Reserves: Prospects and Policy," summary of a workshop at the Harvard John F. Kennedy School of Government, July 2012.
8. "Natural Gas Reserves in the Appalachian Basin Underestimated," energynews24.com, August 29, 2011, http://www.energynews24.com/2011/08/natural-gas-reserves-in-the-appalachian-basin-underestimated.
9. Rick Stouffer, "Marcellus Shale Estimated Natural Gas Yield Rises to Nearly 500 Trillion Cubic Feet," Trib Total Media, July 28, 2009, http://triblive.com/x/pittsburghtrib/business/s_635579.html#axzz2bDgvo0FB.
10. Stephen Moore, "What North Dakota Could Teach California," *Wall Street Journal*, March 11, 2012, http://online.wsj.com/article/SB10001424052970203370604577265773038268282.html.
11. Ibid.
12. Tong, "George Mitchell, 94, Dies."
13. "Unemployment Rates for States," Bureau of Labor Statistics, July 2013, http://www.bls.gov/web/laus/laumstrk.htm.
14. Robert Johnson, "In This Oil Boomtown, Workers with No Experience Are Making $120,000 a Year" Business Insider, March 9, 2012, http://finance.yahoo.com/news/oil-boomtown-workers-no-experience-154600953.html.
15. Catherine Kim and Jessica Hopper, "Now Hiring: North Dakota Oil Boom Creates Thousands of Jobs," *Rock Center with Brian Williams*, October 27, 2011, http://rockcenter.nbcnews.com/_

news/2011/10/27/8495501-now-hiring-north-dakota-oil-boom-creates-thousands-of-jobs?lite.

16. Kris Maher, "Gas Drilling Bringing Jobs to Pennsylvania, but How Many?," Review and Outlook, *Wall Street Journal*, August 2, 2011, http://online.wsj.com/article/SB10001424053111904233404576462543376226516.html?mod=WSJ_hps_sections_news.
17. "A Tale of Two Shale States," *Wall Street Journal*, July 26, 2011, http://online.wsj.com/article/SB10001424052702303678704576442053700739990.html?mod=djemEditorialPage_h.
18. Cullen Browder, "Fracking Brings Risk, Reward to Pennsylvania; Could NC Be Next?," WRAL.com, November 10, 2011, http://www.wral.com/news/local/wral_investigates/story/10359274/.
19. Nicolas Loris, "Hydraulic Fracturing: Critical for Energy Production, Jobs, and Economic Growth," *Backgrounder* no. 2714, Heritage Foundation, August 28, 2012, http://www.heritage.org/research/reports/2012/08/hydraulic-fracturing-critical-for-energy-production-jobs-and-economic-growth.
20. "Unconventional Oil and Gas Production Supports More Than 1.7 Million U.S. Jobs Today; Will Support 3 Million by the End of the Decade, IHS Study Finds," press release, IHS, October 23, 2012, http://press.ihs.com/press-release/commodities-pricing-cost/unconventional-oil-and-gas-production-supports-more-17-millio.
21. "Henry Hub Gulf Coast Natural Gas Spot Price," U.S. Energy Information Administration, August 28, 2013, http://www.eia.gov/dnav/ng/hist/rngwhhdA.htm.
22. Clifford Krauss, "U.S. Inches Toward Goal of Energy Independence," *New York Times*, March 22, 2012, http://www.nytimes.com/2012/03/23/business/energy-environment/inching-toward-energy-independence-in-america.html?pagewanted=all.
23. Elisabeth Rosenthal, "U.S. to Be Top Oil Producer in 5 Years, Report Says," *New York Times*, November 12, 2012, http://www.nytimes.com/2012/11/13/business/energy-environment/report-sees-us-as-top-oil-producer-in-5-years.html?_r=0.
24. Rob Wile, "Prince Alwaleed: Fracking Is Going to Crush the Saudi Economy If Nothing Is Done," Business Insider, July 29, 2013,

http://www.businessinsider.com/saudi-prince-alwaleed-warns-on-fracking-2013-7.

25. Ibid.
26. David G. Savage, "Saudi Arabia–based Charities Still Funding Terrorists, GAO Says," *Washington Times*, September 30, 2009, http://articles.latimes.com/2009/sep/30/nation/na-terror-funding30.
27. Ben Wolfgang, "As N.Y. Fracking Ban Drags On, Leading Energy Company Backs Out," *Washington Times*, August 7, 2013, http://www.washingtontimes.com/news/2013/aug/7/ny-fracking-ban-drags-leading-energy-company-backs/.
28. George Monbiot, "When Will the Oil Run Out?" *Guardian*, December 14, 2008, http://www.theguardian.com/business/2008/dec/15/oil-peak-energy-iea; Sarah Mukherjee, "Warning over Global Oil 'Decline,'" BBC News, October 8, 2009, http://news.bbc.co.uk/2/hi/8296096.stm.
29. Michael Klare, "End of the Petroleum Age?," ed. John Feffer, Foreign Policy in Focus, Institute for Policy Studies, June 26, 2008, http://fpif.org/end_of_the_petroleum_age/.
30. Russell Gold and Ann Davis, "Oil Officials See Limit Looming on Production," *Wall Street Journal*, November 19, 2007.
31. *Energy and Power: A Scientific American Book* (San Francisco: W. H. Freeman, 1971) 39.
32. "U.S. Field Production of Crude Oil," Energy Information Agency, March 15, 2013, http://www.eia.gov/dnav/pet/hist/LeafHandler.ashx?n=PET&s=MCRFPUS2&f=A. Information about the crude reserves can be found in the Energy Information Agency's data tables here: http://www.eia.gov/petroleum/data.cfm#crude.
33. Institute for Energy Research, "The Outer Continental Shelf (OCS)," accessed August 2013, http://www.instituteforenergyresearch.org/issues/ocs/.
34. Institute for Energy Research, *North American Energy Inventory*.
35. "EPCA Phase III Inventory," U.S. Department of the Interior, Bureau of Land Management, last updated May 12, 2011, http://www.blm.gov/wo/st/en/prog/energy/oil_and_gas/EPCA_III.html.

36. Noble Royalties, Inc., "Federal Lands Study," New American Energy Opporunity Foundation, September 2012, http://energyisopportunity.com/wp-content/uploads/2012/09/Federal-Lands-Study.pdf.
37. Ibid.

Chapter Five

1. Lynn O'Shaughnessy, "20 Biggest College Endowments," MoneyWatch, CBS News, February 4, 2013, http://www.cbsnews.com/8301-505145_162-57567214/20-biggest-college-endowments/; "UTIMCO Performance Summary: Preliminary," UTIMCO.org, March 31, 2013, http://utimco.org/extranet/webdata/news/fundperformance.pdf.
2. L.A. Fitzgerald, L. Laurencio, and D. Laurencio, "Geographic Distribution and Habitat Suitability of the Sand Dune Lizard (*Sceloporus arenicolus*) in Texas," paper submitted to Texas Parks and Wildlife Department in fulfillment of requirements on Section 6 project, 2007, 16; L. Laurencio and Fitzgerald, "Atlas of Distribution and Habitat of *Sceloporus arenicolus* in Texas," Texas Cooperative Wildlife Collection, Department of Wildlife and Fisheries Sciences, Texas A&M University.
3. "Request for Emergency Listing of the Sane Dune Lizard (Sceloporus arenicolus) under the Endangered Species Act," WildEarth Guardians, April 9, 2008, http://texasahead.org/texasfirst/esa/task_force/priority/reference_docs/dsl/dsl_wild_earth_guardian_petition.pdf.
4. "Endangered and Threatened Wildlife and Plants; Endangered Status for Dunes Sagebrush Lizard," *Federal Register* 75, no. 39, December 14, 2010, http://www.gpo.gov/fdsys/pkg/FR-2010-12-14/pdf/2010-31140.pdf#page=1.
5. Ibid.
6. Mella McEwen, "Interior Secretary Confident Oil and Gas Development Can Move Forward," Mywesttexas.com, May 11, 2012,

http://www.mywesttexas.com/top_stories/article_ec4c47dd-c970-5ed8-b9ed-73d085af7631.html.

7. "Dunes Sagebrush Lizard," Ecological Services, U.S. Fish and Wildlife Service, http://www.fws.gov/southwest/es/dsl.html.
8. "Endangered and Threatened Wildlife and Plants; Listing the Lesser Prairie-Chicken as a Threatened Species," *Federal Register* 77, 238, December 11, 2012, http://www.gpo.gov/fdsys/pkg/FR-2012-12-11/pdf/2012-29331.pdf.
9. "Lesser Prairie Chicken Interstate Working Group," Western Association of Fish and Wildlife Agencies, accessed August 2013, www.wafwa.org/html/prairie_chicken.shtml.
10. United States of America v. Brigham Oil and Gas, L.P., decisions available at http://www.scribd.com/doc/83119722/US-v-Brigham-Oil-and-Gas-L-P-Bird-kill.
11. Ibid.
12. "EPA Crucify Oil and Gas Companies Mr. Armendariz," YouTube video, posted by "Obliviously Aware," April 26, 2012, http://www.youtube.com/watch?v=WH9ctBMZLxc.
13. Rob Nelson, "The Case against Fracking in *Gasland*," *Village Voice*, September 15, 2010, http://www.villagevoice.com/2010-09-15/film/the-case-against-fracking-in-gasland/full/.
14. "Excerpt: 'Gasland,'" New *York Times* video, clip from *Gasland*, posted June 21, 2010, http://www.nytimes.com/video/2010/06/21/arts/television/1247468091675/excerpt-gasland.html.
15. "'Gasland' Documentary Shows Water That Burns, Toxic Effects of Natural Gas Drilling," AP/Huffington Post, June 21, 2010, updated May 25, 2011, http://www.huffingtonpost.com/2010/06/21/gasland-documentary-shows_n_619840.html.
16. Alyssa Carducci, "'Gasland' Producer Misled Viewers on Lighted Tap Water," Heartland.org, August 1, 2011, http://news.heartland.org/newspaper-article/2011/08/01/gasland-producer-misled-viewers-lighted-tap-water.
17. "'Gasland' Debunked," Energy in Depth Issue Alert, November 2011, http://energyindepth.org/wp-content/uploads/2011/11/Debunking-Gasland.pdf.

18. Mike Hale, "The Costs of Natural Gas, Including Flaming Water," *New York Times*, June 21, 2010, http://www.nytimes.com/2010/06/21/arts/television/21gasland.html.
19. Larry Bell, "'Promise Land's' Fracking Fictions: OPEC Goes Hollywood with Crocudrama," *Forbes*, January 13, 2013, http://www.forbes.com/sites/larrybell/2013/01/13/promised-lands-fracking-fictions-opec-goes-hollywood-with-crocudrama/.
20. Steven and Shyla Lipsky v. Duranter, Carter, Coleman, LLC et al., decision available at http://www.barnettshalenews.com/documents/2012/legal/Court%20Order%20Denial%20of%20Lipsky%20Motion%20to%20Dismiss%20Range%20Counterclaim%202-16-2012.pdf.
21. Bell, "'Promised Land's' Fracking Fictions."
22. Lachlan Markay, "Matt Damon's Anti-Fracking Movie Financed by Oil-Rich Arab Nation," The Foundry (blog), Heritage Foundation, September 28, 2012, http://blog.heritage.org/2012/09/28/matt-damons-anti-fracking-movie-financed-by-oil-rich-arab-nation/.
23. "Wyoming Cleanup Sites," EPA, accessed August 2013, http://www.epa.gov/region8/superfund/wy/pavillion/EPA_ReportOnPavillion_Dec-8-2011.pdf.
24. Chris Tucker, "Update XIII: Six—Actually, Seven—Questions for EPA on Pavillion," Energy in Depth, February 20, 2013, http://energyindepth.org/mtn-states/six-questions-for-epa-on-pavillion-2/; Associated Press, "EPA Implicated Hydraulic Fracturing in Groundwater Pollution at Wyoming Gas Field," Yahoo! News, December 8, 2011, http://news.yahoo.com/epa-implicates-hydraulic-fracturing-groundwater-pollution-wyoming-gas-175134624.html.
25. Kirk Johnson, "EPA Links Tainted Water in Wyoming to Hydraulic Fracturing for Natural Gas," *New York Times*, December 8, 2011, http://www.nytimes.com/2011/12/09/us/epa-says-hydraulic-fracturing-likely-marred-wyoming-water.html.
26. Editorial, "EPA Covers Up the Safety of Fracking," Investors Business Daily, June 21, 2013, http://news.investors.com/

ibd-editorials/062113-661014-epa-rejects-peer-review-of-fracking-study.htm.

27. Associated Press, "EPA Won't Confirm Fracking Pollution Tie, Tells States to Do Their Own Investigation," FoxNews.com, June 21, 2013, http://www.foxnews.com/politics/2013/06/21/epa-wont-confirm-frack-pollution-tie/.
28. "Obama Touts Abound Solar in Weekly Address," YouTube video, President Obama's weekly address from July 3, 2009, posted by "GOPICYMI," June 29, 2012, http://www.youtube.com/watch?v=R86VqQOtTLY.
29. Keenan Steiner, "Another Renewable Energy Loan Recipient Hires Lobbyists, Has Fundraising Ties to Obama," Sunlight Foundation, November 30, 2011, http://reporting.sunlightfoundation.com/2011/another-renewable-grantee-hires-lobbyists-has-fundraising-ties-o/.
30. Ibid.
31. "Bundler: William Kennard," Public Citizen, accessed August 2013, http://www.citizen.org/whitehouseforsale/bundler.cfm?Bundler=13039.
32. "Bundler: Arvia Few," Public Citizen, accessed August 2013, http://www.citizen.org/whitehouseforsale/bundler.cfm?Bundler=77190&mid=20&id=63.
33. C.J. Ciaramella, "Bankrupt Solar Company with Fed Backing Has Cozy Ties to Obama Admin," Daily Caller, September 1, 2011, http://dailycaller.com/2011/09/01/bankrupt-solar-company-with-fed-backing-has-cozy-ties-to-obama-admin/.
34. Michael Bastasch, "Sources, Documents Suggest Government-Subsidized Abound Solar Was Selling Faulty Product," Daily Caller, October 2, 2012, http://dailycaller.com/2012/10/02/sources-documents-suggest-government-subsidized-abound-solar-was-selling-faulty-product/.
35. "The Department of Energy's Disastrous Management of Loan Guarantee Programs," staff report, U.S. House of Representatives Committee on Oversight and Government Reform, March 20, 2012, http://oversight.house.gov/wp-content/uploads/2012/03/FINAL-DOE-Loan-Guarantees-Report.pdf.

36. Michael Sandoval, "DOE-Backed Abound Solar Facing Fraud Investigation in Colorado," The Foundry (blog), Heritage Foundation, October 26, 2012, http://blog.heritage.org/2012/10/26/doe-backed-abound-solar-facing-fraud-investigation-in-colorado/.
37. Sandoval, "Bankrupt Abound Solar to Bury Unused Solar Panels in Cement," The Foundry (blog), Heritage Foundation, February 26, 2013, http://blog.heritage.org/2013/02/26/bankrupt-abound-solar-to-bury-unused-solar-panels/.
38. "Barack Obama's Caucus Speech," transcript of speech given after the Iowa Caucuses, *New York Times*, January 3, 2008, http://www.nytimes.com/2008/01/03/us/politics/03obama-transcript.html?pagewanted=print&_r=0.
39. Lynn Sweet, "Obama Unveils Energy Plan in Detroit. Speech as Prepared. Fact Sheet," *Sun Times*, May 7, 2007, http://blogs.suntimes.com/sweet/2007/05/sweet_blog_special_obama_unvei.html.
40. "Obama energy townhall in Youngstown," transcript of speech in Youngstown, Ohio, Real Clear Politics, August 5, 2008, http://www.realclearpolitics.com/articles/2008/08/obama_energy_townhall_in_young.html.
41. Stephen Moore, "How North Dakota Became Saudi Arabia," *Wall Street Journal*, October 1, 2011, http://online.wsj.com/article/SB10001424052970204226204576602524023932438.html.
42. "Obama's Full Statement at the Stimulus Signing," transcript of remarks given in Denver, CO, *Politico*, February 17, 2009, http://www.politico.com/news/stories/0209/18959.html.
43. Noam Scheiber, *The Escape Artists: How Obama's Team Fumbled the Recovery* (New York: Simon & Schuster, 2012), 103.
44. Daniel Steinberg, Gian Porro, and Marshall Goldberg, "Preliminary Analysis of the Jobs and Economic Impacts of Renewable Energy Projects Supported by the §1603 Treasury Grant Program," National Renewable Energy Laboratory, April 2012, http://www.nrel.gov/docs/fy12osti/52739.pdf; "Recipient Data," Recovery.gov, http://www.recovery.gov/Transparency/RecipientReportedData/Pages/RecipientLanding.aspx.

45. "Our Projects," U.S. Department of Energy Loan Programs Office, accessed August 2013, https://lpo.energy.gov/our-projects/.
46. "The Department of Energy's Disastrous Management of Loan Guarantee Programs," 10.
47. Ibid.
48. "Incentives/Policies for Renewables & Efficiency," Database of State Incentives for Renewables and Efficiency, accessed August 2013, http://www.dsireusa.org/incentives/?State=US.
49. "Renewable Electricity Production Tax Credit (PTC)," Database of State Incentives for Renewables and Efficiency, last updated April 16, 2013, http://www.dsireusa.org/incentives/incentive.cfm?Incentive_Code=US13F&re=1&ee=1.
50. Ibid.
51. "Residential Energy Conservation Subsidy Exclusion (Corporate)," Database of State Incentives for Renewables and Efficiency, last updated July 23, 2012, http://www.dsireusa.org/incentives/incentive.cfm?Incentive_Code=US31F&re=1&ee=1.
52. "Energy-Efficient Commercial Building Tax Deduction," Database of State Incentives for Renewables and Efficiency, last updated October 1, 2012, http://www.dsireusa.org/incentives/incentive.cfm?Incentive_Code=US40F&re=1&ee=1.
53. "USDA—High Energy Cost Grant Program," Database of State Incentives for Renewables and Efficiency, last updated October 11, 2012, http://www.dsireusa.org/incentives/incentive.cfm?Incentive_Code=US56F&re=1&ee=1.
54. "USDA—Repowering Assistance Biorefinery Program," Database of State Incentives for Renewables and Efficiency, last updated January 1, 2013, http://www.dsireusa.org/incentives/incentive.cfm?Incentive_Code=US58F&re=1&ee=1.
55. Ibid.
56. Diane Cardwell and Keith Bradsher, "U.S. Will Place Tariffs on Chinese Solar Panels," *New York Times*, October 10, 2012, http://www.nytimes.com/2012/10/11/business/global/us-sets-tariffs-on-chinese-solar-panels.html?gwh=94F8FEF7A4181A03A33BD1BE6E8209C7.

57. Diane Cardwell, "U.S. Raises Tariffs on Chinese Wind-Turbine Makers," *New York Times*, July 28, 2012, http://www.nytimes.com/2012/10/11/business/global/us-sets-tariffs-on-chinese-solar-panels.html?gwh=94F8FEF7A4181A03A33BD1BE6E8209C7.
58. GAO, "Export-Import Bank: Reaching New Targets For Environmentally Beneficial Exports Presents Major Challenges for Bank," July 15, 2010, http://www.gao.gov/assets/310/307165.html.
59. "Recovery Act: Slow Pace Placing Workers into Jobs Jeopardizes Employment Goals of the Green Jobs Program," Office of Inspector General, Office of Audit, U.S. Department of Labor, September 30, 2011, http://www.recovery.gov/accountability/inspectors/documents/18-11-004-03-390.pdf.
60. "Most States Have Renewable Portfolio Standards," Energy Information Agency, February 3, 2012, http://www.eia.gov/todayinenergy/detail.cfm?id=4850.
61. Robert Bryce, "The High Cost of Renewable-Electricity Mandates," Manhattan Institute's Energy Policy and the Environment Report, February 10, 2012, http://www.manhattan-institute.org/html/eper_10.htm.
62. "Residential Renewable Energy Tax Credit," Database of State Incentives for Renewables and Efficiency, last updated December 11, 2012, http://www.dsireusa.org/incentives/incentive.cfm?Incentive_Code=US37F&re=1&ee=1.
63. "Energy Incentives for Individuals in the American Recover and Reinvestment Act," IRS, last updated May 31, 2013, http://www.irs.gov/uac/Energy-Incentives-for-Individuals-in-the-American-Recovery-and-Reinvestment-Act.
64. "USDA—Repowering Assistance Biorefinery Program."
65. GAO, "Renewable Energy: Federal Agencies Implement Hundreds of Initiatives," February 27, 2012,http://www.gao.gov/products/GAO-12-260.
66. Examiner Watchdog Staff, "Email Includes Energy Department Official's Midnight Confessions on Picking 'Winners and Losers'," *Washington Examiner*, October 26, 2012. Emphasis added.

67. Roberta Rampton and Mark Hosenball, "In Solyndra Note, Summers Said Feds 'Crappy' Investor," Reuters, http://www.reuters.com/article/2011/10/03/us-solyndra-idUSTRE7925C520111003.
68. Larry Summers, Carol Browner and Ron Klain, "Memorandum for the President: Renewable Energy Loan Guarantees And Grants," October 25, 2010.
69. "NRG Energy," CNN Money, accessed August 2013, http://money.cnn.com/magazines/fortune/fortune500/2012/snapshots/11201.html.
70. Statement of Cogentrix CEO Robert Mancini, June 19, 2012, transcript available at http://oversight.house.gov/wp-content/uploads/2012/06/Mancini-Testimony.pdf.
71. Prologis, 10-K, available on the Prologis website at prologis.com.
72. Ormat Technologies, 10-K, available on the Ormat website at ormat.com.
73. Jeff Goodell, "Goodbye, Miami," *Rolling Stone*, June 20, 2013, http://www.rollingstone.com/politics/news/why-the-city-of-miami-is-doomed-to-drown-20130620.
74. James Gleick, *Chaos: Making a New Science* (New York: Penguin, 1988), 18.
75. Ibid., 19.
76. Ibid., 20.
77. Gerald A. Meehl and Thomas F. Stocker, et al., "Quantifying the Range of Climate Change Projections," section 10.5 of *Climate Change 2007: The Physical Science Basis, Contributions of Working Group I to the Fourth Assessment Report of the Intergovernmental Panel on Climate Change* (New York: Cambridge University Press, 2007).
78. Michael Kimmelman, "Going with the Flow," *New York Times*, February 13, 2013, http://www.nytimes.com/2013/02/17/arts/design/flood-control-in-the-netherlands-now-allows-sea-water-in.html?pagewanted=all&_r=0.
79. Garrett A. Lobell, "The Hidden History of New York's Harbor," *Archaeology* 63, no. 6 (November/December 2010), http://archive.archaeology.org/1011/etc/wtc.html.

80. This information came from The Welikia Project (http://welikia.org/m-map.php) and Google Maps.
81. "Filling In below Beacon Hill and Creating the Public Gardens," Bostongeology.com, accessed August 2013, http://bostongeology.com/boston/casestudies/fillingbackbay/fillingbackbay.htm#a.

Chapter Six

1. Amy Levin-Epstein, "Monster Commutes: Should Yours Make the List?," MoneyWatch, CBS News, March 29, 2012, http://www.cbsnews.com/8301-505125_162-57405865/monster-commutes-should-yours-make-the-list/.
2. Jenna Goudreau, "The Cities with the Most Extreme Commutes," Forbes, March 5, 2013, http://www.forbes.com/sites/jennagoudreau/2013/03/05/cities-with-the-most-extreme-commutes/.
3. "Labor Day 2011: September 5," Profile America Facts for Features, United States Census Bureau, August 10, 2011, http://www.census.gov/newsroom/releases/archives/facts_for_features_special_editions/cb11-ff16.html.
4. David Schrank and Tim Lomax, *Urban Mobility Report, 2009*, Texas A&M University Transportation Center for Mobility, Texas Transportation Institute, accessed August 2013, http://americandreamcoalition.org/highways/mobility_report_2009_wappx.pdf.
5. "Motor Vehicle Accidents—Number and Deaths: 1990 to 2009," from the *Statistical Abstract of the United States*, U.S. Census Bureau, accessed August 2013, http://www.census.gov/compendia/statab/2012/tables/12s1103.pdf.
6. Sebastian Thrun et al., "Stanley: The Robot That Won the DARPA Grand Challenge," *Journal of Field Robotics* 23, no. 9 (2006): 661–62, http://www-robotics.usc.edu/~maja/teaching/cs584/papers/thrun-stanley05.pdf.
7. Ibid.

8. Michael Montemerlo et al., "Junior: The Stanford Entry in the Urban Challenge," *Journal of Field Robotics* 25, no. 9 (September 2008): 569–597, available at Stanford.edu, http://robots.stanford.edu/papers/junior08.pdf.
9. "Race Log," Stanford Racing Team, accessed August 2013, http://cs.stanford.edu/group/roadrunner/racelog.html.
10. Sebastian Thrun, "What's Next for Silicon Valley?," *Wall Street Journal*, June 15, 2012, http://online.wsj.com/article/SB10001424052702303807404577434891291657730.html.
11. Thrun, "What We're Driving At," Google: Official Blog, October 9, 2010, http://googleblog.blogspot.com/2010/10/what-were-driving-at.html.
12. Lawrence D. Burns et al., "Transforming Personal Mobility," The Earth Institute, Columbia University, January 27, 2013, http://sustainablemobility.ei.columbia.edu/files/2012/12/Transforming-Personal-Mobility-Jan-27-20132.pdf.
13. Benjamin R. Freed, "Linton Stings Uber After Calling Livery Service 'Illegal,'" DCist.com, January 13, 2012, http://dcist.com/2012/01/linton_stings_uber_leaves_driver_ho.php.
14. Mike DeBonis, "Uber Car Impounded, Driver Ticketed in City Sting," *Washington Post*, January 13, 2012, http://www.washingtonpost.com/blogs/mike-debonis/post/uber-car-impounded-driver-ticketed-in-city-sting/2012/01/13/gIQA4Py3vP_blog.html?hpid=z3.
15. Benjamin R. Freed, "Uber's Fight with D.C. Officials Resumes as Taxi Commission Issues New Sedan Regulations," DCist.com, September 20, 2012, http://dcist.com/2012/09/ubers_fight_with_dc_officials_resum.php; Tony Kalanick, "Strike Down the Minimum Fare Language in the D.C. Uber Amendment," Uber blog, July 9, 2012, http://blog.uber.com/2012/07/09/strike-down-the-minimum-fare/; Julian Hattem and Brendan Sasso, "FTC Sides with Uber in D.C. Taxi Fight," *The Hill*, June 12, 2013, http://thehill.com/blogs/hillicon-valley/technology/305161-ftc-sides-with-uber-in-dc-taxi-fight.

16. Ben Gilbert, "Temporary Restraining Order Blocks Uber and Others from New York City Operations," engadget.com, May 2, 2013, http://www.engadget.com/2013/05/02/uber-hailo-nyc-temporary-restraining-order/.
17. Joe Schoenmann, "Vegas Cab Alternative Runs into Regulations Stifling Competition," *Las Vegas Sun*, March 28, 2012, http://www.lasvegassun.com/news/2012/mar/28/high-tech-limo-service-considers-lv-move-concerned/#axzz2X5Hid3I1.
18. Leena Rao, "Uber Sued by Taxi and Livery Companies in Chicago for Consumer Fraud and More," TechCrunch.com, October 5, 2012, http://techcrunch.com/2012/10/05/uber-sued-by-taxi-and-livery-companies-in-chicago-for-consumer-fraud-and-more/.
19. Rebecca Greenfield, "Cambridge Wants Uber Out, Too," *Atlantic Wire*, August 14, 2012, http://www.theatlanticwire.com/technology/2012/08/cambridge-wants-uber-out-too/55770/.
20. Liz Gannes, "FlightCar Is the Latest Sharing Economy Startup to Face Legal Trouble," AllThingsD.com, June 7, 2013, http://allthingsd.com/20130607/flightcar-is-the-latest-sharing-economy-startup-to-face-legal-trouble/.
21. Mike Ramsey, "Tesla Clashes with Car Dealers," *Wall Street Journal*, June 18, 2013, http://online.wsj.com/article/SB10001424127887324049504578541902814606098.html.

Chapter Seven

1. Greg Klerkx, *Lost in Space: The Fall of NASA and the Dream of a New Space Age* (New York: Random House, 2004), 72.
2. Freeman Dyson, introduction to *The High Frontier: Human Colonies in Space*, by Gerard O'Neill, 3rd ed. (Ontario: Collector's Guide Publishing, 2000), 5.
3. Klerkx, *Lost in Space*, 77.
4. Ibid.; Jeremy Hsu, "Total Cost of NASA's Space Shuttle Program: Nearly $200 Billion," Space.com, April 11, 2011, http://www.space.com/11358-nasa-space-shuttle-program-cost-30-years.html.

5. Klerkx, *Lost in Space*, 44–53.
6. Newt Gingrich, *Window of Opportunity: A Blueprint for the Future* (Tom Doherty Associates, 1987), 59.
7. Clara Moskowitz, "Space Tourist to Unveil Private Mars Voyage Today," Space.com, February 27, 2013, http://www.space.com/19977-private-mars-mission-dennis-tito.html.
8. Robert Zubrin, *The Case for Mars: The Plan to Settle the Red Planet and Why We Must*, rev. ed. (New York: Simon & Schuster, 2011), 282–291.
9. Ibid.

Chapter Eight

1. Tom A. Coburn, *Oklahoma Waste Report: Exposing Washington's Wasteful Spending Habits in Our Own Backyard*, July 2011, http://www.coburn.senate.gov/public/index.cfm?a=Files.Serve&File_id=4f875398-b8bd-4fff-a37a-2cfe984bc3ec.
2. Alex Cameron, "Sidewalks Stir Up Anger over Stimulus in Boynton," News9: Oklahoma's Own, April 12, 2010, http://www.news9.com/story/12298407/sidewalks-stir-up-anger-over-stimulus-in-boynton.
3. Coburn, *Oklahoma Waste Report*.
4. Coburn, *Federal Programs to Die for: American Tax Dollars Sent Six Feet Under*, October 2010, http://www.coburn.senate.gov/public//index.cfm?a=Files.Serve&File_id=406062be-b959-4798-b55e-39029197a9fc.
5. Coburn, *Subsidies of the Rich and Famous*, November 2011, http://www.coburn.senate.gov/public//index.cfm?a=Files.Serve&File_id=544ae3e7-195b-40ad-aa84-334fdd6a5e1f.
6. Coburn, *Wastebook 2012*, March 2013, http://www.coburn.senate.gov/public//index.cfm?a=Files.Serve&File_id=b7b23f66-2d60-4d5a-8bc5-8522c7e1a40e.
7. Ibid.

8. Coburn, *2008: Worst Waste of the Year*, December 2008, http://www.coburn.senate.gov/public/index.cfm?a=Files.Serve&File_id=4dd719bc-8177-4772-93a6-926146d420d7.
9. Ibid.
10. Ibid.
11. Coburn, *Wastebook 2012*.
12. Ibid.
13. Ibid.
14. Ibid.
15. Ibid.
16. Coburn, *2011 Wastebook: A Guide to Some of the Most Wasteful and Low Priority Government Spending of 2011*, December 2011, http://www.coburn.senate.gov/public//index.cfm?a=Files.Serve&File_id=2b11ca9d-1315-4a7c-b58b-e06686e3aece.
17. Ibid.
18. Ibid.
19. Coburn, *Wastebook 2010: A Guide to Some of the Most Wasteful Government Spending of 2010*, December 2010, http://www.coburn.senate.gov/public//index.cfm?a=Files.Serve&File_id=4a184ddb-cd85-4052-b38b-5a1116acca8c.
20. Ibid.
21. Ibid.
22. Ibid.
23. Ibid.
24. Ibid.
25. Coburn, *Wastebook 2012*.
26. Phillip Swarts, "GAO Comptroller Says Federal Government Is Flying Blind, Wasting Money," *Washington Times*, July 10, 2013, http://www.washingtontimes.com/news/2013/jul/10/gao-comptroller-says-federal-government-is-flying-/?page=all; "Financial Aid Performance Management: More Reliable and Complete Information Needed to Address Federal Management and Fiscal Challenges," statement of Gene L. Dodaro, testimony before the Committee on Oversight and Government Reform, House of Rep-

resentatives, GAO.gov, July 10, 2013, http://gao.gov/assets/660/655803.pdf.

27. Ronald Bailey, "Federal Regulations Have Made You 75 Percent Poorer," *Reason*, June 21, 2013, http://reason.com/archives/2013/06/21/federal-regulations-have-made-you-75-per/print.
28. Clyde Wayne Crews Jr., *2013 Ten Thousand Commandments: An Annual Snapshot of the Federal Regulatory State*, Competitive Enterprise Institute, May 2013, http://cei.org/sites/default/files/Wayne%20Crews%20-%2010,000%20Commandments%202013.pdf.
29. Ibid.
30. Peter Schroeder, "Dodd-Frank Regulations Would Fill 28 Copies of 'War and Peace,'" *The Hill*, July 19, 2013, http://thehill.com/blogs/regwatch/finance/312205-dodd-frank-regs-dwarf-war-and-peace.
31. Crews, *2013 Ten Thousand Commandments*.
32. Ibid.
33. Ibid.
34. Nicole V. Crain and W. Mark Crain, *The Impact of Regulatory Costs on Small Firms*, Small Business Administration Office of Advocacy, September 2010, http://www.sba.gov/sites/default/files/The%20Impact%20of%20Regulatory%20Costs%20on%20Small%20Firms%20(Full).pdf.
35. Crews, *2013 Ten Thousand Commandments*..
36. John W. Dawson and John J. Seater, "Federal Regulation and Aggregate Economic Growth," *Journal of Economic Growth* 18, no. 2 (June 2013): 137–77, available at NCSU.edu, http://www4.ncsu.edu/~jjseater/regulationandgrowth.pdf.
37. Bailey, "Federal Regulations Have Made You 75 Percent Poorer."
38. Alex Tabarrok, Launching the Innovation Renaissance: A New Path to Bring Smart Ideas to the Market Fast (New York: TED Books, 2011).
39. Meredith Whitney, *Fate of the States: The New Geography of American Prosperity* (New York: Penguin Group, 2013), 89–90.
40. Ibid., 60.

41. Eden Martin, "Even After Proposed Pension Fix, Illinois Would Still Be Broke," *Chicago Sun-Times*, April 17, 2013, http://www.suntimes.com/news/martin/19543712-452/after-pension-fix-illinois-still-broke.html.
42. "Report: California's Actual Debt At Least $848B; Could Pass $1.1T," CBS Sacramento, May 1, 2013, http://sacramento.cbslocal.com/2013/05/01/report-californias-actual-debt-set-at-848b-could-pass-1-1t/.
43. Walter Russell Mead, "California Rail Fail: Captain Brown and the Great White Train," *American Interest*, January 15, 2012, http://blogs.the-american-interest.com/wrm/2012/01/15/california-rail-fail-captain-brown-and-the-great-white-train/.
44. "Population of the 100 Largest Urban Places: 1950," U.S. Bureau of the Census, June 15, 1998, http://www.census.gov/population/www/documentation/twps0027/tab18.txt; "2010 Population Finder," U.S. Census Bureau, 2010 Census, http://www.census.gov/popfinder/?fl=26:2600380:2622000.
45. Tyler Durden, "Detroit by the Numbers," ZeroHedge.Com, July 24, 2013, http://www.zerohedge.com/news/2013-07-24/detroit-numbers.
46. Jeff Green and Mark Clothier, "U.S. Automakers Thrive as Detroit Goes Bankrupt," Bloomberg.com, July 19, 2013, http://www.bloomberg.com/news/2013-07-19/u-s-automakers-thrive-as-detroit-goes-bankrupt.html.
47. Bill Nojay, "Lessons from a Front-Row Seat for Detroit's Dysfunction," *Wall Street Journal*, July 29, 2013, http://online.wsj.com/article/SB10001424127887323829104578623422748612116.html.
48. Mark Steyn, "The Downfall of Detroit," National Review Online, July 19, 2013, http://www.nationalreview.com/article/353959/downfall-detroit-mark-steyn.
49. "Detroit Imposes Business Hours at Police Precincts," FoxNews.Com, January 12, 2012, http://www.foxnews.com/us/2012/01/12/detroit-imposes-business-hours-at-police-precincts/.
50. Steyn, "The Downfall of Detroit."

51. Ibid.
52. Dennis Cauchon, "Some Federal Workers More Likely to Die Than Lose Jobs," *USA Today*, July 19, 2011, http://usatoday30.usatoday.com/news/washington/2011-07-18-fderal-job-security_n.htm.
53. David Ignatius, "The Benghazi E-mails' Backside-Covering," *Washington Post*, May 17, 2013, http://www.washingtonpost.com/opinions/david-ignatius-the-benghazi-e-mails-backside-covering/2013/05/17/8a4f6fa2-be62-11e2-97d4-a479289a31f9_story.html.
54. Josh Hicks, "IRS Issued Billions in Improper Refunds, Report Says," *Washington Post*, April 24, 2013, http://www.washingtonpost.com/blogs/federal-eye/wp/2013/04/24/irs-issued-billions-in-improper-refunds-report-says/.
55. David A. Fahrenthold, "Watch Him Pull a USDA-Mandated Rabbit Disaster Plan Out of His Hat," *Washington Post*, July 16, 2013, http://www.washingtonpost.com/politics/watch-him-pull-a-usda-mandated-rabbit-disaster-plan-out-of-his-hat/2013/07/16/816f2f66-ed66-11e2-8163-2c7021381a75_story.html.
56. David A. Fahrenthold, "One Grower's Grapes of Wrath," *Washington Post*, July 7, 2013, http://www.washingtonpost.com/lifestyle/style/one-growers-grapes-of-wrath/2013/07/07/ebebcfd8-e380-11e2-80eb-3145e2994a55_story.html.
57. Rob Nikolewski, "IRS Went After 83-Year-Old Tea Party Granny," *Washington Examiner*, May 2013, http://washingtonexaminer.com/irs-went-after-83-year-old-tea-party-granny/article/2530131; Joe Newby, "GOP Rep. Aaron Schock: IRS Asked about Content of Pro-Life Group's Prayers," Examiner.com, May 18, 2013, http://www.examiner.com/article/gop-rep-aaron-schock-irs-aked-about-content-of-pro-life-group-s-prayers.
58. Gregory Korte, "IRS Approved Liberal Groups While Tea Party in Limbo," *USA Today*, May 15, 2013, http://www.usatoday.com/story/news/politics/2013/05/14/irs-tea-party-progressive-groups/2158831/.
59. James Bovard, "A Brief History of IRS Political Targeting," *Wall Street Journal*, May 14, 2013, http://online.wsj.com/article/SB10001424127887324715704578482823301630836.html.

60. Joseph Weber, "EPA Acknowledges Releasing Personal Details on Farmers, Senator Slams Agency," FoxNews.Com, April 9, 2013, http://www.foxnews.com/politics/2013/04/09/epa-acknowledges-giving-out-personal-info-in-request-that-included-data-on/.
61. Josh Hicks, "IRS Released Confidential Info on Conservative Groups to ProPublica," *Washington Post*, May 14, 2013, http://www.washingtonpost.com/blogs/federal-eye/wp/2013/05/14/irs-released-confidential-info-on-conservative-groups-to-propublica/.
62. Sarah Kliff, "Budget Request Denied, Sebelius Turns to Health Executives to Finance Obamacare," *Washington Post*, May 10, 2013, http://www.washingtonpost.com/blogs/wonkblog/wp/2013/05/10/budget-request-denied-sebelius-turns-to-health-executives-to-finance-obamacare/.

Chapter Nine

1. Alexis de Tocqueville, *Democracy in America*.
2. Franklin D. Roosevelt, "Second Inaugural Address," January 20, 1937, transcript available at Bartleby.com, http://www.bartleby.com/124/pres50.html.
3. Amity Shlaes, *The Forgotten Man: A New History of the Great Depression* (New York: HarperCollins, 2007), 230.
4. Ibid., 151.
5. "The FarmVille 2 Almanac of Records and Facts," *Zynga* (blog), January 4, 2013, http://blog.zynga.com/2013/01/04/farmville2 infographic/.
6. Gavin Newsom and Lisa Dickey, *Citizenville: How to Take the Town Square Digital and Reinvent Government* (New York: Penguin Group, 2013).
7. Ibid.
8. Ibid.
9. Ibid.
10. Lincoln's December 1862 message to Congress captured him at his very best, and his words seem especially appropriate for this time:

"We can succeed only by concert. It is not 'can any of us imagine better?' but, 'can we all do better?' The dogmas of the quiet past, are inadequate to the stormy present. The occasion is piled high with difficulty, and we must rise—with the occasion. As our case is new, so we must think anew, and act anew. We must disenthrall ourselves, and then we shall save our country."

11. Newsom and Dickey, *Citizenville*, emphasis added.
12. Randall Stross, "Wearing a Badge, and a Video Camera," *New York Times*, April 6, 2013, http://www.nytimes.com/2013/04/07/business/wearable-video-cameras-for-police-officers.html?pagewanted=all&_r=0.
13. Adam Mazmanian, "Issa Reboots DATA Act," FCW.com, May 16, 2013, http://fcw.com/articles/2013/05/16/issa-data-act.aspx.
14. Newsom and Dickey, *Citizenville*.
15. Ibid.
16. Tim O'Reilly, "Gov 2.0: It's All about the Platform," TechCrunch, September 4, 2009, http://techcrunch.com/2009/09/04/gov-20-its-all-about-the-platform/.
17. Statistics come from the U.S. Government Printing Office and Bureau of Economic Analysis. They are available here: http://stats.areppim.com/stats/stats_usxrecxspendxgdp.htm.

Chapter Ten

1. Karen Weise, "Record U.S. Poverty Rate Holds as Inequality Grows," *Bloomberg Businessweek*, September 12, 2012, http://www.businessweek.com/articles/2012-09-12/record-u-dot-s-dot-poverty-rate-holds-as-inequality-grows; Hope Yen, "U.S. Poverty on Track to Rise to Highest since 1960s," Huffington Post, July 22, 2012, http://www.huffingtonpost.com/2012/07/22/us-poverty-level-1960s_n_1692744.html.
2. National Poverty Center, "Poverty in the United States: Frequently Asked Questions," University of Michigan, accessed August 2013, http://www.npc.umich.edu/poverty/.

3. These data comes from World Bank statistics last updated July 12, 2013, using Google Public Data, http://www.google.com/public-data/explore?ds=d5bncppjof8f9_&ctype=l&met_y=ny_gdp_mktp_kd_zg#!ctype=l&strail=false&bcs=d&nselm=h&met_y=ny_gdp_mktp_kd_zg&scale_y=lin&ind_y=false&rdim=region&idim=country:USA&ifdim=region&hl=en_US&dl=en_US&ind=fals.
4. Joseph Dalaker, *Poverty in the United States: 2000*, U.S. Census Bureau, Current Populations Reports (U.S. Government Printing Office: Washington, D.C.), September 2001, http://www.census.gov/prod/2001pubs/p60-214.pdf.
5. Ibid.
6. Derek Thompson, "The Economic History of the World (After Jesus) in Four Slides," *Atlantic*, June 20, 2012, http://finance.yahoo.com/news/economic-history-world-jesus-4-190149386.html.
7. Deirdre McCloskey, "Growth, Quality, Happiness, and the Poor," quoted in John Tomasi, *Free Market Fairness* (Princeton, NJ: Princeton University, 2012), 59.
8. National Poverty Center, "Poverty in the United States: Frequently Asked Questions."
9. Associated Press, "Big Cities Battle Dismal Graduation Rates," CBS News, February 11, 2009, http://www.cbsnews.com/2100-201_162-3985714.html.
10. Lisa Fleisher and Alison Fox, "State Officials Release New Test Score Data, Proficiency Drops," *Wall Street Journal*, August 7, 2013, http://blogs.wsj.com/metropolis/2013/08/07/state-officials-release-new-test-score-data-scores-drop/.
11. Terence P. Jeffrey, "U.S. Department of Education: 79% of Chicago 8th Graders Not Proficient in Reading," CNSNews.com, September 10, 2012, http://cnsnews.com/news/article/us-department-education-79-chicago-8th-graders-not-proficient-reading.
12. "Report: Nearly Half of Detroiters Can't Read," CBS Detroit, May 4, 2011, http://detroit.cbslocal.com/2011/05/04/report-nearly-half-of-detroiters-cant-read/.

13. "Salman Khan Talk at TED 2011," YouTube video, from Salman Khan's talk at TED 2011, posted by "khanacademy," March 9, 2011, www.youtube.com/watch?v=gM95HHI4gLk.
14. See, for example, the Office of Policy Development and Research, *Evaluation of the Family Self-Sufficiency Program: Prospective Study*, U.S. Department of Housing and Urban Development, February 2011.
15. Stephanie Wynn, "Once Upon a Time, a Little Girl Was Paid to Read," *Wall Street Journal*, November 12, 1990.
16. Michael D. Tanner, "Welfare: A Better Deal Than Work," National Review Online, August 21, 2013, available at Cato.org, http://www.cato.org/publications/commentary/welfare-better-deal-work.
17. "Poverty," U.S. Census Bureau, accessed August 2013, http://www.census.gov/hhes/www/poverty/data/.
18. Charles M. Blow, "Marriage and Minorities," *New York Times*, August 2, 2013, http://www.nytimes.com/2013/08/03/opinion/blow-marriage-and-minorities.html?_r=0.
19. Ibid.
20. *One in 31: The Long Reach of American Corrections*, Pew Center on the States, March 2009, http://www.pewstates.org/uploadedFiles/PCS_Assets/2009/PSPP_1in31_report_FINAL_WEB_3-26-09.pdf.
21. "Start Early, Finish Strong: How to Help Every Child Become a Reader," U.S. Department of Education, July 1999, http://www2.ed.gov/pubs/startearly/intro.html.
22. "Ensuring Success for Young Children: Early Childhood Literacy," Association of Small Foundations, November 2008, http://www.aecf.org/~/media/Pubs/Topics/Education/Other/EnsuringSuccessforYoungChildrenLiteracy/Ensuring_Success_Early_Literacy.pdf.
23. Cindy Hendricks, James E. Hendricks, and Susie Kauffman, "Literacy, Criminal Activity, and Recidivism," American Reading Forum, http://www.americanreadingforum.org/yearbook/yearbooks/01_yearbook/html/12_Hendricks.htm.

24. "Reentry Trends in the U.S.: Recidivism," Bureau of Justice Statistics, accessed August 2013, http://www.bjs.gov/content/reentry/recidivism.cfm.
25. National Gang Intelligence Center, *2011 National Gang Threat Assessment: Emerging Trends*, FBI, http://www.fbi.gov/stats-services/publications/2011-national-gang-threat-assessment.
26. Congressional Record, 13238, June 11, 1986.
27. John Gallagher, "Downtown Detroit Projects Slowed by Financing, Red Tape," *Detroit Free Press*, May 10, 2013, http://www.freep.com/article/20130510/BUSINESS06/305100017/Detroit-development-Gallagher-Midtown.

Chapter Twelve

1. Nancy Cordes, "Probe: IRS Contractor Won Up To $500 Million in Questionable Bids," CBS News, June 25, 2013, http://www.cbsnews.com/8301-18563_162-57591030/probe-irs-contractor-won-up-to-$500-million-in-questionable-bids/.
2. The video of this exchange can be seen in Ashley Killough, "Duckworth Rips Witness over Veterans Disability Claim," Political Ticker, CNN, June 26, 2013, http://politicalticker.blogs.cnn.com/2013/06/26/duckworth-rips-witness-over-veterans-disability-claim/.
3. Chana Joffe-Walt, "Unfit for Work: The Startling Rise of Disability in America," *All Things Considered*, National Public Radio, March 2013, report available online at http://apps.npr.org/unfit-for-work/
4. Ibid.
5. Luke Rosiak, "EXography: Many Disability Recipients Admit They Could Work," *Washington Examiner*, July 30, 2013, http://washingtonexaminer.com/exography-many-disability-recipients-admit-they-could-work/article/2533626.
6. Joffe-Walt, "Unfit for Work."
7. Ibid.
8. Rosiak, "EXography: Many Disability Recipients Admit They Could Work."

INDEX

$2.50 a Gallon: Why Obama Is Wrong and Cheap Gas is Possible Now, 82
60 Minutes, 126–27

A

Abigail Alliance for Better Access to Developmental Drugs, 3, 68
Abound Solar, 94–95, 99
Academy and College of Philadelphia, 154
Adventures of Huckleberry Finn, The, 38
Affordable Care Act, 55. *See also* Obamacare
African Americans, 39, 180–81
Agora Cyber Charter School, 36–37
Airbnb, 163
Alaskan Wildlife Refuge (ANWR), 82
Albuquerque, 158
Aldrin, Buzz, 128
Allen, Paul, 132
Al Qaeda, 79
Alzheimer's disease, 13, 134, 188–89, 191
Amazon, 8, 40, 141, 161
American Philosophical Society, 154
Americans
  and commuting, 108
  and disabilities, 195–99
  and the economy, 14, 141
  and energy policy, 14, 75–78, 84, 90, 93–95
  and environmental policy, 95–98
  and government breakdown, 143, 152–53, 167–68, 207–8
  and healthcare policy, 50, 56, 61–70, 187–90
  and incarceration, 181–83
  and outer space, 127–29
  and poverty, 169–75, 179–81, 185–86
  and previous breakouts, 6–8, 156–61
  role in breakout of, 6, 19–23, 78, 84, 205, 207–9
American Spectator, 56
Amsterdam, 103

Android devices, 9, 39, 175
Angry Birds, 37
Ann Arbor, MI, 115
Annie E. Casey Foundation, 182
Ansari X Prize, 132–33
AOL, 8
APIs, 163
Apollo 13, 46
Apollo (space missions), 125–26
Apple, 8, 10, 162
Archaeology, 104
Armendariz, Alfredo, 91
Armstrong, Neil, 128
Army Materiel Command, 166–67
Around the World in Eighty Days, 7–8
asthma, 49
Atala, Anthony, 61–66, 68
Atlanta, GA, 108
AT&T, 9
Audi, 112–13, 123
automobiles, 7–8, 14, 18, 108–9, 112, 116, 122–23
Avis, 120
A&W Root Beer, 116

B

Bakken Shale, 74–77, 81, 96
Ballou High School, 173–74
Bangalore, 10
Barnett Formation (Barnett Shale), 73–74
Bayer, 68
Beacon Hill, MA, 104
Beaver County, PA, 78
Ben-Amara, Ridah, 118–19
Benghazi attacks, 146
Best Buy, 161
"big government," 155–56, 202
bin Talal, Alwaleed (prince), 79
Bloomberg, 44
Bloomberg, Michael, 183–84
Blow, Charles M., 181
Blue Diamond (nut company), 69–70
BMW, 112–13
Boston, 104, 108, 120
Bowker, Kent, 74
Boynton, OK, 137–38
Branson, Richard, 132
Bratton, William, 147, 183–84
breakout
  in cures, 187–92
  in disabilities, 193–99
  in education, 26–48
  in energy, 13–14, 71–84
  in government, 15, 153–168
  in health, 13, 49–70, 187, 192–93
  in materials and manufacturing, 12–13
  opposition to, 5–6, 16–19, 57, 85–105, 117–18, 121–24, 127–31, 141, 152, 165–66, 189–90, 192, 207–9
  the past and, 6–8, 11, 18–19, 103, 154,
  "personal" breakout, 71
  possibility for, 5–6, 8–16, 19–23, 71–72, 133–34, 141, 201–9
  from poverty, 169–186
  in space, 125–33
  in transportation, 14–15, 107–24
breakout champions, 22–23, 69, 201–6
breakthroughs, 72, 191–92, 201, 203–5
  as buildup to breakout, 5–6, 9, 16–17, 20, 134–35, 171
  in education 20, 173, 182, 199
  in energy, 14, 20, 72–81
  in government, 152
  green energy and the lack of, 98
  in health, 2, 4, 13, 20, 50–52, 58, 61, 63–66, 68–69, 188–92, 199, 208
  NASA and, 128
  obstacles to, 202–3, 208–9
  in the penal system, 173, 181–83
  potential for, 15–16
  in poverty, 173
  the prize system and, 191
  in public safety, 183–86
  in space, 133
  in transportation, 15, 45, 108, 112
  in welfare policy, 173, 176–78
Brigham Oil and Gas, 88
Brin, Sergey, 20
Brown, Jerry, 126
bureaucracy
  corruption in the, 15, 151–53
  in education, 26–27, 48, 174–75
  in environmental policy, 87–88

in health, 50–51, 55–56, 58–60, 66, 192
NASA and, 128–32
as obstacle to breakout, 1, 12, 18, 20, 22, 27, 66–67, 72–73, 95, 118–19, 143, 146–49, 152, 156, 159–65, 185, 191, 198, 201–8
replacement of the, 159–66, 201–6, 208
size of the, 139–40, 148–49, 155, 159–61
in transportation, 122–23
waste in the, 139, 153
in welfare, 178–79
Bureau of Ocean Energy Management, 82–83
Burns, Lawrence, 115
Burroughs, Abigail, 1–5, 64, 68–69
Bush, George W., 127–28

C

C225, 3. *See also* Erbitux
California, 15, 46, 98, 125–26, 131, 148, 155–56, 161
debt of, 142–43
self-driving cars and, 14, 45, 111–12, 117
California Faculty Association, 47
Cambridge, MA, 120
cancer
Abigail Burroughs and, 1–2
cost of, 29
cures for, 13, 51, 53, 64, 66, 108, 134, 188
treatments for, 2–4, 68, 192
candles, 7, 16–18, 152
Cantor, Eric, 159
Carnegie Mellon, 110–11, 114
Carter, Jimmy, 72, 74–75
Case for Mars, The, 133–34
Castillo, Braulio, 193–95
Caterpillar, 14, 114, 166
Cato Institute, the, 179
Cavalier Daily, 1
CBS News, 108
Center for Health Transformation, 147
Chaos: Making a New Science, 101–2
Chicago, IL, 39, 120, 158, 174, 184
Children's Hospital of Philadelphia, 59
China, Chinese, 12, 15, 21, 97–98, 127, 166, 190
Chronicle of Higher Education, 33, 47
Chrysler, 94, 121, 123–24
Churchill, Winston, 203
Chu, Steven, 96
Citizen Cosponsors, 159
Citizenville, 156, 162
Citizenville, 155–58, 161–65
civil service system, the, 146, 165–66
Cleveland Clinic, 50–51
Club of Rome, 102
CNN, 183
Coburn, Tom, 138–39, 160–61, 164
Code of Federal Regulations (*CFR*), 140
Cogentrix, 100
CollegePlus, 25
Collins, Chris, 167
Colorado, 74, 88, 90–91, 94
Colson, Charles, 181
Columbia, 128
Columbia University, 44, 115
commutes, commuting, 107–8, 113, 115, 117
Competitive Enterprise Institute, 140
CompStat, 147, 183–84
Congressional Budget Office, 57
Constitution, the, 165, 168
Continental Resources, 74, 76, 88
Contra Costa County, CA, 142
Cornell University, 44
corrections systems, 172, 181–83
Council Hill, OK, 138
Creative Destruction of Medicine, The, 52
Crossfire, 183
Curiosity (Mars rover), 128
Curious George, 38

D

Daily Caller, 95
Daily Mail, 131
DARPA Grand Challenge, 109–11
DARPA Urban Challenge, 110–12
data.gov, 161
Dean, Howard, 160
Defense Advanced Research Projects Agency (DARPA), 109–12

Deficit Free American Summit, 166
Deming, W. Edwards, 166
Democrats, 94, 155–56, 162, 193
Detroit Department of Transportation, 144
Detroit Free Press, 185
Detroit, MI, 7–8, 143–45, 174, 179, 184–85
diabetes
  cost of, 49, 63, 188
  cures for, 13, 51, 61–62, 66, 134
  disability and, 196
Digimorph.com, 58
Digital Accountability and Transparency Act (DATA), 162
Disneyland, 116
District of Columbia, the, 179. *See also* Washington, D.C.
District of Columbia Taxicab Commission (DCTC), 119–21
Dodd-Frank Act, 140
Dragon (spacecraft), 133
Drill Here, Drill Now, Pay Less, 80
Drucker, Peter, 174
Duckworth, Tammy, 176, 193–97
dunes sagebrush lizard (sand dune lizard), 85–88
Duolingo, 175
Dutch, the, 78, 103
Dyson, Freeman, 127, 129

E

Earning by Learning, 177–78
Earth Institute (Columbia University), 115
eBay, 163
EcoFinder, 159
Edison, Thomas, 7, 17–18, 171
Egypt, 15, 159
Ehrlich, Paul, 102
Einstein, Albert, 171
Elk Lake School District, 78
Energy in Depth, 91–93
Energy Information Agency, 75–76, 78
English Channel, the, 131
environmentalists, 18, 20, 84, 86, 91–93, 148
Environmental Protection Agency, 89, 91–93, 148–51, 168
Environmental Protection Agency Region 6, 89
Erbitux, 3–4
Escape Artists, The, 96
Europe, 6–7, 21, 113, 122, 153, 170, 190
Export-Import Bank, 98

F

Facebook, 9, 26, 48, 116, 156, 159, 161–62, 205
Facemash, 8–9
Falcon 1 rocket, 133
Falcon 9 rocket, 133
Falls Church, VA, 1
FarmVille, 156–58
Fate of the States, 142
Federal Aviation Administration (FAA), 66–67
Federal Bureau of Investigation (FBI), 184
federal government
  breakdown of the, 143–53
  breakout and, 6, 8, 10–11, 15–16, 132, 153–68, 187, 190, 201–9
  bureaucracy and, 22, 48, 51–52, 155–6, 162–66, 201
  corruption in the, 87–89, 95, 98–100, 198
  disability and, 193–96, 198–99
  poverty and the, 175, 178–79
  as prison guard of the past, 9, 48, 52, 54, 57, 80, 82–83, 103, 122–24, 127–31, 189, 193, 207
  reliance on the, 8, 143, 159–60
  state governments and, 15
  subsidies for green technology, 96–100
  waste and the, 134, 137–41
Federal Register, 140
FedEx, 114, 141
Ferrara, Peter, 178, 180–81
Few, Arvia, 94–95
Finch, Atticus, 37
Fiorina, Carly, xi–xii
First Amendment, the, 151
Fish and Wildlife Service, 86–88, 93
FlightCar, 120
Florida, 117–18, 138
Florida's Virtual Academies (FLVA), 36
Food and Drug Administration (FDA), 2–4, 20, 50, 55–58, 64–70, 188–90
Ford, 108, 122–23, 141

Ford F150, 122–23
Forgotten Man, The, 155
for-profit education companies, 39, 42, 44
Fort Worth, TX, 73
fossil fuels, 14, 90, 93, 100
Founding Fathers, 150, 164
Fox, Josh, 90–91
FrackNation, 91
Frank, Barney, 184
Franklin, Benjamin, 154
Freedom of Information Act, 161
Frogue, Jim, 147

G

gangs, 172–73, 184
Gasland, 89–91
Gasland II, 92
Gates, Bill, 10, 20, 27, 32–33, 46–47, 103
Genesis probe, 130
Geneva Convention on Road Traffic, 118
George Group, the, 166
George, Mike, 166–67
Georgia Institute of Technology (Georgia Tech), 47
Giuliani, Rudy, 147, 183–84
Gleick, James, 101–2
Glenn, John, 128
Golden Spike, 133
Goldman Sachs, 100
Google, 8–11, 14, 26, 39, 45–46, 109, 111–12, 117, 140,
Google Lunar X Prize, 132
Google Maps, 9, 107, 161
Google Now, 163
Google Street View, 45, 111
Gore, Al, 101
"Gov 2.0," 162, 165
Government Accountability Office (GAO), 139
GPS, 67, 109, 118, 120, 122, 159, 163, 183
Grassley, Chuck, 149
Great Depression, the, 8, 97
Great Recession, the, 181
"Great Society," the, 179
Greece, Greek, 30, 73
Groupon, 163
Gulf of Mexico, 76

H

"H1ghlander" (self-driving car), 110
Hahne, Marty, 147–48
Hailo, 120
Hale County, AL, 195–97
Hamm, Harold, 74–76, 84, 96, 140
Harshal, 30
Hart-Rudman Commission on National Security, 143
Harvard Kennedy School of Government, 143
Harvard Medical School, 39
Harvard University, 8–9, 39, 43–44, 85
Hearst, William Randolph, 131
Henninger, Daniel, 67
Heritage Foundation, the, 178–81
Hertz, 120
High Frontier, The, 125–27
Hope in the Unseen, A, 173
horizontal drilling, 14, 74, 81. *See also* fracking
Houston, TX, 108, 158
Hubbert, M. King, 72, 81
Hudson, Henry, 104
Hudson River, the, 104
Huffington Post, the, 33, 90
Hurricane Katrina, 143
hydraulic fracturing, 14, 78, 92–93. *See also* fracking

I

IBM, 9–10
Ignatius, David, 146
IHOP, 138
IHS Global Insight, 178
Illinois, 142–43, 193
Inconvenient Truth, An, 101
Independence Hall, 154
Independent Payment Advisory Board (IPAB), 56
Independent Petroleum Association of America, 91
Industrial Revolution, the, 170
"innobucks," 158
Inside Higher Ed, 47
Institute for Energy Research, 75
Institute for Policy Studies, 81

Insurance Institute for Highway Safety, 122–23
internal combustion engine, 7–8, 112, 203
Internal Revenue Service (IRS), 50, 129, 147, 150–51
International Space Station, 127, 130, 132–33
iPads, 11, 26, 52, 161, 175
iPhones, 9, 11, 38, 52, 100, 107, 118, 155, 159, 162–63, 175

J

Japan, 170, 190
Jennings, Cedric, 173, 176
Jobs, Steve, 10, 20, 100, 183
Joffe-Walt, Chana, 195–97
Johns Hopkins Hospital, 2
Johnson, Nancy, 148
Jones, Van, 183
Josephson Institute of Ethics, 151
Journal of Economic Growth, 141

K

K12, 37
Kai Ani, Karim, 33–34
Kaplan, 39–44, 182, 199
Kelly, Raymond, 184
Kemp, Jack, 184–85
Kennard, Lydia, 94
Kennedy, John F., 170
Khan Academy, 27–35, 43, 140, 175, 177, 182–83
Khan, Salman, 27–38, 44–45, 157, 175
Kickstarter, 163
Kindles, 26, 66
Klerkx, Greg, 126, 129
Komatsu, 114

L

Lance, Mike, 137–38
Las Vegas, 120
Las Vegas Strip, the, 17
Latinos, 172
Lean Six Sigma, 166–68
"learning coaches," 37–38, 177
learning science, 39–42, 175, 209
Left, the, 6, 17, 19, 92–93, 95, 100, 103, 208
lesser prairie chicken, 88
Library of Congress, 161
LIDAR sensors, 111
light bulbs, 7, 18–19, 152
Limits to Growth, The, 102
Lincoln, Abraham, 17, 159, 171
Lindbergh, Charles, 8, 131
Linton, Ron, 119–21
Lipitor, 54
Lockheed Martin, 130
Los Altos, CA, 29–30
Losing Ground, 178
Lost in Space, 126, 130
Louisiana, 74
LSAT, the, 40
Lunar Lion (Penn State), 132

M

Machiavelli, Niccolò, 19, 21
Mad Max, 144–45
malaria, 13
Manhattan Institute, the, 67
Manhattan, NY, 103–5, 115
Manor Labs, 157–58,
Manor, TX, 157–58
Marcellus Shale, 76–77, 79
Mars, 133–34, 138
Mars climate orbiter, 130
Massachusetts Institute of Technology (MIT), 28, 39
"massively open online courses (MOOCs), 46–47
Mathis, Otis, 145
McAleer, Phelim, 90–91
McCloskey, Deirdre, 171
McCrea, Jim, 99
McDonald's, 56, 116
"Mechanical Turk" platform, the, 40
Medicaid, 50, 147, 179, 181, 187–88
Medicare, 50–51, 56, 147, 187–88, 196
Mellow Mushroom, 138
Mercury 6 (space mission), 128
Microsoft, 8, 10, 132, 141, 176

Middle East, the, 15, 92, 109
Migratory Bird Act, 88
Minerals Management Service, 76, 82
Mir (Russian space station), 130
Mitchell Energy, 73–74
Mitchell, George, 73–75, 80, 140
Mojave Desert, 109
moon, the, 21, 125–29, 132–33
Moore, Stephen, 76, 96
Morocco, 138
motion pictures, 7, 92
Motorola, 8 166
Moynihan, Daniel Patrick, 178
Murnaghan, Sarah, 58–61, 63
Murray, Charles, 178
Musk, Elon, 120, 133

N

NASA, 127–34, 138
National Academy of Engineering, 167
National Highway Transportation Safety Administration (NHTSA), 121–24
National Institute for Literacy, 182
National Institute of Medicine, 167
National Institutes of Health, the, 191
National Institute of Standards and Technology, 120
National Mall, the, 165
National Public Radio, 195
National Recovery Administration (NRA), 155
Naval Warfare Systems Center, 167
NBC News, 52, 77
Netflix, 42
Netherlands, the, 103
New Deal, the, 155
Newfield Production Company, 88
Newsom, Gavin, 15, 155–60, 162–64
Newton, Isaac, 28
NewtUniversity.com, 39
New York, 74, 76, 80, 90, 104, 116, 143, 167, 183
New York City, NY, 8, 17, 108, 120, 131, 134, 145, 147, 174, 184
New York Times, the, 90, 93, 126, 181
Nexavar, 68
Nike FuelBand, 57
Nissan, 112
Nixon, Richard, 150
Noble Royalties, 83–84
Noble, Scott, 84
Nojay, Bill, 144
Nolan, Pat, 181
North American Energy Inventory, 75
North Dakota, 71, 74, 76–77, 81, 88
Norton, Robert, 121–24
Norvig, Peter, 45–46
NRG Energy, 94, 100

O

Obama administration, the, 94, 99, 133, 146, 152, 161
Obama, Barack, 127, 185
  Abound Solar and, 94
  campaign of, 39
  debt and, 147
  green energy and, 96–97, 99–100
  higher education and, 48
  IRS scandal and, 150
  reelection of, 95
  unemployment rate and, 77
Obamacare, 50–51, 55–58, 60, 151–52, 172, 185, 187
Ohio, 74, 76–77
Olasky, Marvin, 176, 178
O'Neill, Gerard, 125–27, 133
"Open 311," 158–59
Opportunity (Mars rover), 128
O'Reilly, Tim, 162–64
Organization of Petroleum Exporting Countries (OPEC), 79, 96
Organ Procurement and Transplantation Network (OPTN), 58–60
Ormat Technologies, 100
Orteig Prize, 131
Orteig, Raymond,
Oshkosh Defense, 109, 114
Outer Continental Shelf (U.S.), 82
Oxford University, 39

P

Page, Larry, 20, 111
"paper NASA," the, 129

Parkinson's disease, 189
Pathfinder (spacecraft), 130
Pavillion, WY, 92
PBS, 43
"peak oil," 13–14, 72–74, 79, 81, 84, 95
"peak oil theory," 72, 84
Peltzman, Sam, 67
Pennsylvania, 36, 74, 76–78, 80, 91–92
Pennsylvania Department of Labor, 78
Pentagon, the, 143, 166
Permian Basin, 85
Permian Basin Petroleum Association, 87
personalized medicine, 13, 51–52, 209
Pierce, Caitlin, 25
pioneers of the future, 20, 22, 140, 152, 191, 201–6, 208
  in education, 11, 27, 175
  in energy, 72, 74, 76, 84, 96
  in health, 60–61, 199
  in outer space, 131–32
  in transportation, 113, 121
Population Bomb, The, 102
Prince, The, 19
Princeton University, 44, 125
prison guards of the past, 17, 20–23, 26–27, 32, 35, 81, 127, 139, 141, 152, 163, 165, 201, 205–6, 208–9
  conventional wisdom as a, 72
  in education, 27, 32, 35, 38, 48
  in energy, 80–81, 84, 140
  in environmental policy, 88–89, 92–93, 95–100, 105
  in healthcare, 51, 55, 56, 69–70, 189
  NASA as a, 127–33
  in poverty, 174, 178–79, 184
  in transportation, 117–24
prize model, the, 131
Prologis, 94, 100
Promised Land, 92
Prudhoe Bay, AK, 76
PT Cruiser, 123–24
Purdon, Timothy, 88

Q

Quantum Reservoir Impact, 77

R

radios, 7, 37, 94
"real NASA," the, 129
Reason, 141
Rector, Robert, 180
regenerative medicine, 61, 64–66, 69, 188–90, 209
Republicans, 159
Reuters, 57, 99
Rialto, CA, 161
Ricardo UK, 113
Richmond, VA, 107, 158
Right, the, 6, 17, 19,
Rio Tinto, 114
Robotics Institute (Carnegie Mellon), 114
Rocky Mountains, 116
Rolling Stone, 100–1
Roosevelt, Franklin Delano, 154
Rotterdam, 103
Rowe, Tim, 57
Roy, Avik, 67
rule of law, 145, 150–52
Russia, Russians, 127–28, 130
Rust Belt, 77

S

Sacramento, CA, 143
Sagan, Carl, 126
Saint Louis Zoo, 138
Salazar, Ken, 94
Saleri, Nansen, 77
San Diego, CA, 143
San Francisco, CA, 108, 120, 156, 158
San Jose State University, 46–47
SARTRE project, 113
Saudi Arabia, 78–79
Saxberg, Bror, 39–43, 45, 175
Say's phoebe, 88
Scheiber, Noam, 96
Scripps Translational Science Institute, 52
SeeClickFix, 158
self-driving cars, 14, 45, 109, 112–18, 121, 124, 140, 171, 199, 209
Shepherds Flat, 100
Shepperd, Ben, 87
Shetty, Devi Prasad, 50

Shlaes, Amity, 155
SideCar, 120
Siri, 38
Slate, 33
Small Business Administration, 140
smartphones, 10, 54, 57, 115, 118, 121, 158, 162, 175–76, 203
Social Security Administration, 195–96, 198
Social Security Disability Insurance, 18
solar power, 126
South Carolina, 138, 167
Southern Methodist University, 91
Soyuz capsules, 127
Space Adventures, 133
SpaceShipOne, 132
SpaceShipTwo, 132
SpaceX, 133
Spirit (Mars rover), 128
Stanford University, 45, 48, 109–11
"Stanley" (self-driving car), 109–110
Staples, 12
steam engine, the, 6, 9
Stevens, Bill, 74
Stop Paying the Crooks, 147
Street View (Google), 45, 111
Stryker, Pat, 94, 99
suburbs, 116, 120, 157, 185
Summers, Lawrence, 99–100
Sundance Film Festival, 90
Suskind, Ron, 173
Susquehanna County, PA, 78
Sutent, 68
sweat-equity model, 184

T

Tabarrok, Alex, 141
Taliban, 79
teachers' unions, 20, 27, 35, 38, 42, 93
Tea Party, 150–51
TED conference (2011), 27
TED conferences, 112
TED talks, 29, 62
Temporary Assistance to Needy Families, 181
Tenth Amendment, 165, 168
TerraMax, 114
Tesla, 120–21
Texas A&M Department of Wildlife and Fisheries Sciences, 86
Texas A&M University, 85–86
Theranos, xii
THOMAS, 161
three-dimensional (3-D) printers, 12, 62–63
Timberlake, Perry, 197, 199
Times Square (NY), 17
Tito, Dennis, 132
Tocqueville, Alexis de, 153–54, 159, 164
Topol, Eric, 51–57, 68–69, 190
Toyota, 111–12
Tragedy of American Compassion, The, 176, 178
transcontinental railroad, the, 6
Trippi, Joe, 160
tuberculosis, 13
Tufts University Center for the Study of Drug Development, 67
Turco, Andy, 71, 74, 77
Turkey, 15, 156
Twitter, 11, 33, 158, 163, 205

U

Uber, 118–21, 140
Udacity, 46–48, 109, 140, 176, 182, 199
Udall, Morris, 126
underground transit systems, 8
Unfit for Work, 195, 198
United Arab Emirates, 92
United Kingdom Energy Research Centre, 80
University of Chicago, 67
University of Pennsylvania, 154
University of Phoenix, 44
University of Texas at Austin (UT Austin), 57, 85
University of Virginia, 1, 3
UPS, 114
U.S. Air Force, 163
U.S. Army, 166–67, 193, 203
USA Today, 146

U.S. Congress, 55, 139, 149, 155, 159, 161, 190, 193, 204
FDA and, 67
Kathleen Sebelius and, 151
members of, 21, 167, 184, 192
NASA and, 130, 132–33
natural gas terminals and, 80
Organ Procurement and Transplantation Network (OPTN) and, 58
Washington, D.C., and, 107
U.S. Department of Agriculture, 147
U.S. Department of Defense, 98, 165–66
U.S. Department of Energy, 94
U.S. Department of Health and Human Services (HHS), 56, 58–59
U.S. Department of Justice, 88–89, 165
U.S. Department of State, 165
U.S. Department of Veterans Affairs, ix–xii
U.S. House Energy and Commerce Committee, 57
U.S. House Oversight Committee, 162, 193
U.S. Interior Department, 83, 86, 94
U.S. Justice Department, 88–89, 93, 151, 165
U.S. Military Academy prep school, 193
U.S. Supreme Court, 3, 155
U.S. Treasury Department, 97

V

VanOverbeke, Lisa, 185
Verne, Jules, 7–8
Village Voice, 90
Virgin Galactic, 132
virtual charter schools, 36–38. *See also* Agora Cyber Charter School; Florida's Virtual Academies (FLVA)
Visa, 161
Volvo, 112–13, 122
von Eschenbach, Andy, 188
von Neumann, John, 101

W

Wahhabism, 79
Wall Street Journal, the, 45, 67, 76, 81, 96, 120, 144, 151, 173, 177
Washington, D.C., 20, 50, 75, 80, 107–8, 118–19, 137–38, 143, 149, 160, 164–66, 173, 185
Washington Examiner, 196, 198
Washington Post, 3, 33, 35, 119, 146–47
"Web 2.0," 162
WebMD, 57, 161
welfare, 20, 89, 152, 173, 178–81, 185, 196–97
Western Europe, 170
West Virginia, 76–77
White House, the, 94, 99, 155
Whitney, Meredith, 142
Wikipedia, 9, 26, 54, 92
Williams, Billy Dee, 138–39
Williston, ND, 71–72, 77
Window of Opportunity, 131
wind power, 98
Wired, 33
"wireless medicine," 54–55
World Trade Center, 103
World War II, 18, 116, 151
World of Warcraft, 139
Wright brothers, the, 8

X

X Prize Foundation, 132–33
X-rays, 8, 64

Y

Yahoo!, 8
Yale University, 44
YouCut, 159
YouTube, 11, 26–28, 32, 35, 162

Z

Zipcars, 115
Zubrin, Robert, 133–34
Zuckerberg, Mark, 8
Zynga, 156